ASTEROIDS

A History

ASTEROIDS

A History

CURTIS PEEBLES

SMITHSONIAN INSTITUTION PRESS
WASHINGTON AND LONDON

EDITOR: Lorraine Atherton
PRODUCTION EDITOR: Ruth W. Spiegel
DESIGNER: Amber Frid-Jimenez

Library of Congress Cataloging-in-Publication Data
Peebles, Curtis.
Asteroids : a history / Curtis Peebles
p. cm.
Includes bibliographical references and index.
ISBN 1-56098-389-2 (alk. paper)
1. Asteroids. I. Title.
QB651.P44 2000
523.44—dc21
00-020733

BRITISH LIBRARY CATALOGUING-IN-PUBLICATION DATA AVAILABLE
Manufactured in the United States of America
06 05 04 03 02 01 00 5 4 3 2 1

♾ The recycled paper used in this publication meets the minimum requirements of the American National Standard for Permanence of Paper for Printed Library Materials Z39.48-1984.

CONTENTS

ACKNOWLEDGMENTS

First and foremost, my thanks go to three pioneers of astronomy—Eugene M. Shoemaker, Carolyn Shoemaker, and David H. Levy—for taking the time to talk with me. The enthusiasm they felt for astronomy and the joy they took in their work was clearly evident. This made Eugene Shoemaker's death in an accident all the more tragic. I also wish to thank Brian G. Marsden, of the Minor Planet Center, and Jennifer Palm, Postmaster of Palomar Mountain. Thanks also go to Lee Saegesser and the NASA History Office for providing copies of their asteroid files. Most of the sources used in this book would not have been available to me except for their efforts. Thanks also to the Palomar Mountain Lodge.

Insights into the San Diego streetlighting controversy were provided by Michael Anderson, as well as several other individuals. I also thank my fellow members of the Save Palomar Committee and the many hundreds of people who signed our petitions and sent letters to the San Diego City Council. I also wish to express my admiration for the late David Lasser, for his work on the controversy and for serving as an example of a man with the courage of his convictions. Finally, my thanks to the late Eric Blair. Through his writings, I was able to understand the greater meaning of that struggle.

ACKNOWLEDGMENTS

INTRODUCTION

LOST AMID THE STARS, there are mountains in the sky. Some are worlds in their own right, others are the irregular splinters of collisions ages ago, while still others are merely orbiting boulders. Although most are found between the orbits of the planets Mars and Jupiter, some travel the inner solar system. Others orbit the outer rim of the solar system. Together they are the asteroids, and this is the history of their discovery and what we have come to know of them.

Two centuries ago humans did not know asteroids even existed. Human ideas about the asteroids have changed significantly in that time. In the early nineteenth century they were viewed as equals of the major planets. By the end of that century and for most of the twentieth century, they were ignored as "vermin." Astronomers had begun looking away from the solar system, toward the farthest reaches of space. They had no time for places closer to home. It has been only in the last three decades that interest in them has revived. Today we know their mineral composition, we have imaged their surfaces by radar and orbiting telescope, and we have flown space probes past them. Some show signs of having undergone volcanic flows, while others contain water and ice. We have come to understand that asteroids provide a glimpse into the origins of the solar system itself. There has also been the realization that asteroids are intertwined with both our distant past and possible future. We exist because of an impact 65 million years ago, and asteroids carry the potential of ending human existence.

In the past several years, this aspect has attracted increased scientific and public attention. Search programs have been started, and controversy has flared up over what to do about the possibility that an asteroid or comet could strike Earth. The ideas of impacts and planetary defense have also entered popular culture. Two of the most successful movies of 1998 were *Deep Impact* and *Armageddon*. Both dealt with efforts to deflect objects about to hit Earth. What might come as a surprise to most moviegoers is that the first study of the technology to prevent an asteroid impact was made in 1967, and the concept of planetary defense originated in the early nineteenth century, with one of the most important figures in English literature.

The history of the asteroids is a very human endeavor, involving careful research and intuition, advanced technology and the simplest of telescopes, some of the past two centuries' most important scientists as well as an amateur astronomer observing from a window above a Paris café, and newspaper headlines and carefully hidden Cold War secrets. Asteroid studies also show how science has changed. Today they are no longer the sole province of astronomers but also of geologists, biologists, engineers, and weapons designers. In the story of the asteroids, we can see case studies of who makes scientific discoveries, of how discoveries are made, of how science works, and how decisions really get made. It is also the story of personal feuds and political ignorance, and of bitter controversies lasting decades.

Here is how the story began, and where it has led us.

1

DISCOVERY OF THE ASTEROIDS

Then felt I like some watcher of the skies
When a new planet swims into his ken.

John Keats, "On First Looking into Chapman's Homer," 1817

BEFORE A DISCOVERY CAN be made, human imagination must be opened to new possibilities. So it was with the discovery of the asteroids. In the cramped, Earth-centered universe of the Greeks, there was no room for the possibility of unknown planets. For more than a thousand years after their incorporation into Christian theology, the ideas of Ptolemy held absolute sway.

In 1543 Nicolaus Copernicus made the first challenge to the Earth-centered universe. Copernicus proposed that all the planets of the solar system orbited around the Sun. One consequence of this arrangement was that the orbits of the planets now had sizable gaps between them. Copernicus's ideas came at a time when Europe was shaking off the mental chains of the Dark Ages. In 1492, a half-century before, Columbus had sailed to the New World, opening the first great age of exploration. At home, the religious reformation, led by such figures as Martin Luther, challenged the Catholic Church's monopoly on ideas.

Tycho Brahe was one of those who saw the new possibilities in astronomy. Tycho's research involved the exact measurement of the movements of the planets. Over more than twenty years, Tycho had made thousands of precise, systematic measurements. The data were used by Tycho's assistant, Johannes Kepler, to determine the actual shape of the planets' orbits. Kepler found, counter to the belief going back to Ptolemy, that the orbits were not perfect circles but were actually ellipses. Kepler derived the basic laws of orbital motion, but as yet the process by which it occurred was a mystery.[1]

One of Kepler's discoveries was that the orbit of Jupiter was very much farther away than that of Mars—there was a huge gap between the two planets. This offended Kepler's sense of proportion. Kepler wrote in 1596, "Between Jupiter and Mars I place a planet."[2]

At this same time, the telescope was invented, representing the great technological breakthrough of Renaissance astronomy. Starting in 1609, Galileo began making telescopic observations. He discovered the craters of the Moon, the phases of Venus, sunspots, and the moons of Jupiter. For the Catholic Church, these discoveries were heresy. In 1616 the Roman Inquisition determined: "That the doctrine that the Sun is the center of the world and immoveable was false and absurd, formally heretical and contrary to Scripture, whereas the doctrine that the Earth was not the center of the world but moved, and has further a daily motion was philosophically false and absurd and theologically at least erroneous."

Galileo was warned not to "hold, teach, or defend" Copernicus's ideas. He persisted, and in 1633 he was brought before the Inquisition. Under threat of torture, Galileo was forced to denounce his own beliefs. He spent the remainder of his life under house arrest. Galileo, now blind and helpless, died in 1642, still deemed a threat.

Isaac Newton was born on Christmas Day 1642, the same year Galileo died. During the plague years of 1665–1666, Newton first turned his attention to a theory of gravitation. It was not until 1687, however, that his discoveries were published. Newton showed that gravity was the result of the mutual attraction between two masses. This attraction decreased in an inverse-square relationship. That is, if the distance between two objects is doubled, the force acting on them will be one-fourth as strong. What Kepler had shown empirically, Newton could now explain theoretically. More important, gravitation was universal—it applied equally to an apple falling from a tree, to the Sun and a planet, and to the farthest reaches of space.

THE TITIUS-BODE LAW

The events that led to the discovery of the asteroids began in 1772 with the publication of the second German-language edition of Charles Bonnet's book *Reflection on Nature*. The book's translator, Johann Daniel Titius, a professor of mathematics at Wittenberg, added a footnote:

> Divide the distance from the Sun to Saturn into 100 parts; then Mercury is separated by 4 such parts from the Sun; Venus by 4 + 3 = 7 such parts; the Earth by 4 + 6 = 10; Mars by 4 + 12 = 16. But notice that from Mars to Jupiter there comes a deviation from this exact progression. After Mars, there follows a distance of 4 + 24 = 28 parts, but so far no planet or satellite has been sighted there. . . . Let us assume that this space without doubt belongs to the still undiscovered satellites of Mars. . . . Next to this for us still unexplored space there rises Jupiter's sphere of influence at 4 + 48 = 52 parts; and that of Saturn at 4 + 96 = 100.

That same year, nearly the exact same wording appeared in Johann Elert Bode's book *Introduction to the Study of the Starry Sky*. Bode had lifted Titius's footnote without giving him credit. Bode said, "Can one believe that the Founder of the Universe had left this space empty? Certainly not." Both books were reprinted during the 1770s. As yet, the idea of a planet between Mars and Jupiter was based on religious grounds.[3]

On March 13, 1781, William Herschel discovered the planet Uranus. Bode (who was first to suggest the name) noted that the progression predicted a planet at a distance of 196 (4 + 192 = 196). The actual distance was 192. It was also not until this monograph that Bode admitted it was Titius who first came up with the idea.[4] Astronomers saw the Titius-Bode Law, as it subsequently became known, as having been proven by Uranus. The quest for unknown planets had begun. As the eighteenth century neared its close, attention focused on the gap between Mars and Jupiter. Here the Titius-Bode Law predicted a planet. The hunt for the missing planet was under way, and the police were hot on its trail.

THE CELESTIAL POLICE

Baron Franz Xaver von Zach was among those fascinated by the possibility of discovering this unknown planet. In 1783 he met and observed with William Herschel. It is presumed that they discussed the Titius-Bode Law and the missing planet. Two years later Zach made a rough estimate of the unknown planet's orbit. He lacked the critical element of where on its orbit the planet was located, however. In 1791 he became the director of the Seeberg Observatory near Gotha, in what is now Germany. In 1798 he started *Geographical Ephemeris,* the first scholarly astronomical journal. Although these duties occupied most of his time, Zach also spent several years in an unsuccessful search for the planet.

The task was daunting; without an indication of the object's location, each star visible in the telescope would have to be checked against a star chart. Any star not on the chart would have to be checked the next night; if it had moved, then it was the unknown planet. The charts were often inaccurate, adding to the number of possibilities to be rechecked. Zach eventually realized that a more organized approach was necessary. In August 1798 Zach held a meeting at Gotha with Bode and the French astronomer Joseph Lalande. He proposed a cooperative effort involving several astronomers searching for the missing planet.

In 1800 Zach turned over the editorship of *Geographical Ephemeris* to others and founded a less formal journal called *Monthly Correspondence.* With the lightened workload, Zach was able to turn his attention fully to the missing planet. On September 11, 1800, Zach held a meeting at the Lilienthal Observatory near Bremen. On hand were Zach, Johann Schroeter (the observatory director), Karl Harding (Schroeter's assistant), Heinrich Olbers, Ferdinand Adolph von Ende, and Johann Gildemeister. Neither Lalande nor any other French astronomer was involved, as the Hanover government had banned all contact after the French Revolution. The plan was simple. As all the known planets orbited along the ecliptic (the projection of Earth's orbit against the sky), it was assumed that the unknown planet would as well. The ecliptic would be divided into 24 sections, 15 degrees wide and 7 or 8 degrees north and south. Each astronomer would search his assigned sector while making a new star chart. Schroeter was selected as president of the group, and Zach was its secretary. They dubbed themselves the Celestial Police. Letters were sent to astronomers across Europe inviting them to join the search.[5]

GIUSEPPE PIAZZI AND THE DISCOVERY OF CERES

One of the astronomers Zach selected for the Celestial Police was Giuseppe Piazzi, an Italian monk with a history of independent thought and run-ins with his religious superiors. Between 1769 and 1779, Piazzi's ideas had cost him jobs as a philosophy professor at Genoa, Malta, and Ravenna. It was not until 1780 that he was able to find a position at the Academy of Palermo in Sicily. During the late 1780s, Piazzi traveled to England and, like Zach, met Sir William Herschel. Piazzi's observing session with Herschel came to a bad end—Piazzi fell off a ladder and broke his arm. In 1790 Piazzi became the director of the Palermo Observatory. During the next decade he was involved with the updating of a star catalog by the French astronomer Nicolas-Louis de Lacaille. The original catalog was filled with errors, so he had to check each star one by one.[6]

On the night of January 1, 1801, the first night of the nineteenth century, he was looking for a star in the constellation of Taurus. Piazzi found the star, then noted another dim star close by that was not on his chart. When he checked again the next night, the dim star had moved. He continued to observe the object on the nights of January 3 and 4. Bad weather intervened, and it was not until January 10 and 11 that he was able to recover the object. After additional observations on January 13, 14, and 17 through 23, Piazzi wrote a letter to his friend Barnaba Oriani on January 24, 1801. By this time, Piazzi had begun to realize what he had discovered: "I have announced the star as a comet. But the fact that the star is not accompanied by any nebulosity and that its movement is very slow and rather uniform has caused me many times to seriously consider that perhaps it might be something better than a comet. I would be very careful, however, about making this conjecture public."

The following day Piazzi wrote a similar letter to Bode. He continued to observe the object until February 11, 1801, when poor weather and a severe illness brought Piazzi's work to a halt. Bode did not receive Piazzi's letter until March 20. (Napoleon's invasion of Italy had begun as Piazzi was making his observations.) Bode immediately concluded that the missing planet predicted by the theory had been found. The discovery was also announced in Zach's *Monthly Correspondence* during the summer of 1801.

Ironically, having been discovered after years of searching, the missing planet was again missing. The object's orbital motion had, by this time, carried it into the evening twilight, and so it could not be observed again until the early fall of 1801. The observations Piazzi had made in January and early February were not sufficient to calculate an orbit. William Herschel and others searched unsuccessfully for the object, and it was feared the object had been permanently lost.

Among the subscribers to *Monthly Correspondence* was Carl Friedrich Gauss, one of the most brilliant mathematicians ever to live. In September 1801 he was thinking about a new method of orbital calculation that would require only a few positions. Then he learned of Piazzi's discovery. Gauss wrote later: "Nowhere in the annals of astronomy do we meet with so great an opportunity, and a greater one could hardly be imagined, for showing most strikingly, the value of this problem, that in this crisis and urgent necessity, when all hopes of discovering in the heavens this planetary atom, among innumerable small stars after the lapse of nearly a year, rested solely upon a sufficiently approximate knowledge of its orbit to be based upon these very few observations."

The work was done in October 1801, and Gauss sent the results to Zach. On December 7, 1801, Zach observed four stars at the location predicted by Gauss. Due to bad weather, he was not able to look again until December 18. He found one of the stars was missing. It was not, however, until the early morning hours of January 1, 1802, that Zach was finally able to confirm he had recovered the object. A few hours later, Heinrich Olbers independently spotted the object. It was a year to the day since it was first seen by man.[7]

The next matter was to name the new planet. The interest was great, and as will be detailed later, there were several different proposals. Finally, the name Ceres, after the Roman goddess of corn and harvests, as well as the patron goddess of Sicily, was accepted.[8]

For those involved with the search for the unknown planet, rewards were soon in coming. On behalf of the king of Sicily, the secretary of state wrote Piazzi commending him on the discovery of Ceres. Originally, the king wanted to present Piazzi with a medal. Instead, the king agreed to the more practical suggestion of a new instrument for the Palermo Observatory. Fame also had its dark side for Piazzi. Critics would taunt him by saying, "Piazzi was discovered by Ceres."

For Zach there was the satisfaction of recovering Ceres, as well as knowing the valuable role *Monthly Correspondence* had played. Without the channel of communications the journal had provided, Ceres may well have been lost for much longer. Gauss further refined the mathematical procedures he first used to recover Ceres. When the work was completed in 1809, it became the standard procedure for orbital calculations.

With the discovery of Ceres, it seemed that the solar system was complete. However, there were more surprises awaiting in the darkness.

PALLAS

On March 28, 1802, Olbers was looking for Ceres when he noticed another moving object nearby. The second missing planet was named Pallas, after the Greek goddess of wisdom, war, and the liberal arts. When Gauss calculated its orbit, the results were unexpected. Although Ceres and Pallas had nearly identical 4.6-year periods of revolution, the orbit of Pallas was tilted over 34 degrees. The high inclination meant Pallas spent only a small fraction of its orbit within the ecliptic. The search procedure of the Celestial Police (of which Olbers was a charter member) would not have worked. Had it not been for the happenstance that Pallas was passing near Ceres when Olbers was making his observations, it would have been many decades before someone stumbled across Pallas.[9]

On April 1, 1802, Sir William Herschel observed both Ceres and Pallas. With one eye he looked at each object through the telescope, while the other eye looked at a small, distant disk. By comparing them with the disk, which had a known size and distance, Herschel was able to make a crude size estimate. It had been expected that the new objects would be the size of the other planets.[10] Herschel's results came as a surprise. He derived values of 259 kilometers for Ceres and 236 kilometers for Pallas (considerably below the modern values of around 940 kilometers and 588 kilometers).

It was also Herschel who coined the name for the new class of objects. Ceres and Pallas both "resemble small stars so much as hardly to be distinguished from them. From this, their asteroidal appearance, if I may use that expression, therefore I shall take my name, and call them Asteroids." The term "asteroid," Latin for starlike, was later joined by "minor planet" and "planetoid."

Olbers was the first to put forward a theory of asteroid origin. In a letter to Herschel, he wrote, "Could it be that Ceres and Pallas are just a pair of fragments, or portions of a once greater planet which at one time occupied its proper place between Mars and Jupiter, and was in size more analogous to the other planets, and perhaps millions of years ago, either through the impact of a comet, or from an internal explosion, burst into pieces?" Olbers concluded that there were a great many asteroids. The Celestial Police decided to continue the search. Directly counter to this was an 1807 suggestion that the asteroids were formed as small bodies between Mars and Jupiter. This theory also presumed that there were other asteroids awaiting discovery. One estimate was at least ten more.

JUNO AND VESTA

It was not until two years after Pallas was discovered that the next asteroid was found. Karl Harding (another of the Celestial Police) spotted it on September 1, 1804. The object was named Juno, after the wife of Jupiter. As with Pallas, Juno had its own peculiarities. The modern value for its estimated size is only 248 kilometers, much smaller than either Ceres or Pallas.

The fourth asteroid was first observed on March 29, 1807, by Olbers (the first person to discover two asteroids), as a result of his theory that the asteroids were the fragments of an exploded planet. Olbers reasoned that while the fragments would be propelled into differing orbits by the blast, the orbits themselves would all still intersect at the point of the explosion and at a location 180 degrees away. Olbers believed that these two points in space were in the northwest part of the constellation Virgo and the western part of Cetus. This simplified the search—Olbers could wait for the asteroids to come to him.[11]

Olbers became so familiar with the stars in these two regions that he was able to spot the asteroid at a glance, even before its motion was apparent. Olbers took measurements over several nights and sent them to Gauss. Within ten hours of receiving them, Gauss had calculated the asteroid's orbit. (Even with a pocket calculator that would be difficult.) As Gauss had calculated the asteroid's orbit, he was given the honor of naming it. He selected "Vesta," after the patroness of the vestal virgins, goddess of fire, and sister of Ceres.

Vesta has its own unusual properties. It is the brightest asteroid; under ideal conditions it will reach a magnitude of 6.0 or 5.9 for several weeks.

This is just bright enough to be seen by the naked eye. The first to see it without a telescope was Johann Schroeter. He said later, "We both afterwards saw this planet several times, with our naked eyes, when the sky was clear, and when it was surrounded by smaller invisible stars, which precluded all possibility of mistaking it for another. This proves how very like the intense light of this planet is to that of a fixed star."[12]

Although it orbits slightly closer to Earth than Ceres, Vesta is only about 576 kilometers in diameter, a little more than half the estimated size of Ceres and slightly smaller than Pallas. The reason for Vesta's brightness is that its surface reflects 23 percent of the sunlight striking it. This is more than four times the light reflected by Ceres and most other asteroids. Vesta also appears pinkish to some observers. Because of their sizes, Ceres, Pallas, and Vesta are now known as the Big Three asteroids.[13]

The discovery of Vesta brought to a close the first era of asteroid research. The Celestial Police disbanded, leaving Olbers to continue the search alone. He finally gave up in 1816, having decided the effort was futile. No more asteroids were found between 1807 and 1845. During those 38 years, all but one of the individuals connected with the discovery of the first four asteroids died.

Johann Schroeter was the first. The final years of his life had been tragic. In 1806 Lilienthal came under the control of the French, who did not pay Schroeter for his work as chief magistrate of the town. Schroeter's financial situation became worse in 1810 when the French authorities dismissed him from his post. The worst blow came in 1813, following Napoleon's disastrous invasion of Russia. A retreating French detachment skirmished with a small group of Cossacks near Lilienthal. A wounded French officer reported they had been fired on by local peasants. The French retaliated by burning Lilienthal. Destroyed in the fire was the government building where Schroeter kept his books and observations. Several days later, French troops looted his observatory, where the Celestial Police had held their first meeting. With most of his life's work destroyed, Schroeter died in August 1816.[14]

He was followed in death by Herschel in 1822, Bode and Piazzi in 1826, Zach in 1832, Harding in 1834, and Olbers in 1840. Only Gauss, who lived until 1855, still remained when the fifth asteroid was found.

2

VERMIN OF THE SKIES

Their supervision requires an unduly prodigious amount of time. In fact even now the value of a new discovery is hardly in proper proportion to the added work it causes.

Rudolph Wolf

DESPITE THE EFFORTS OF Olbers and the expectation of more discoveries, no other asteroids were found for nearly four decades. In 1830 Friedrich Wilhelm Bessel urged German observatories to undertake a sky-mapping project to assist in the search for new asteroids. Yet success continued to elude astronomers.

THE SECOND PHASE OF DISCOVERY

The drought was finally ended by Karl L. Hencke. Rather than a professional astronomer, Hencke was a postmaster in Driesen, Germany. He had resumed the search in 1830 and spent the next fifteen years, alone and unaided, plotting the locations of the stars, looking for a moving point of light. On the night of December 8, 1845, Hencke was looking for Vesta when he accidentally discovered the fifth asteroid. It was given the name Astraea, after the goddess of justice. Like Olbers, Hencke continued to search, and on July 1, 1847, he was rewarded with the sixth asteroid, later named Hebe (the Greek goddess of youth and cupbearer to the other gods). The name was given by Gauss, who alone among the first generation of asteroid researchers had lived to see the beginning of the second phase.

The year 1847 saw two more discoveries: Iris on August 13 and Flora on October 18. Both were first spotted by J. R. Hind. Metis, the ninth asteroid, was discovered in 1848; Hygiea in 1849; and Parthenope, Victoria, and Egeria in 1850.[1] It is important to realize that many of the early asteroid discoverers were amateur astronomers, often using rather modest equipment. This pattern would continue into the early years of the twentieth century.

One of these early asteroid discoverers was Hermann Goldschmidt. He was a German-born artist living in Paris, France. He came to astronomy late in life and quite by chance. When Goldschmidt was 45, he attended a lecture on astronomy by Urbain-Jean-Joseph Leverrier. A brilliant mathematician, Leverrier was famed for his work in celestial mechanics, having done the calculations that led to the discovery of the planet Neptune. Goldschmidt was enthused by the lecture, and with money from the sale of a painting of Galileo, he bought a 2-inch telescope. Goldschmidt's first discovery was Lutetia (the Latin name for Paris), on November 15, 1852.[2] He continued his search without success for nearly two years before spotting Pomona on October 26, 1854. Atalante followed a year later on October 5, 1855, and 1856 saw two discoveries, Harmonia on March 31 and Daphne on May 22.

The year 1857 was the high point of Goldschmidt's career. It began with the discovery of Nysa on May 27, Eugenia a month later on June 27, and Melete on September 9. Ten days later, Goldschmidt made history. On the night of September 19, he became the first to discover two asteroids in a single night, Doris and Pales. The final years of his life saw another four discoveries: Europa and Alexandra on February 4 and September 10, 1858; Danae on September 9, 1860; and his fourteenth and final asteroid, Panopaea, on May 5, 1861. What is all the more remarkable is that some of his discoveries were made from his window above the Café Procope.

It was another eighteen years before a similar double discovery was made, this time by C. H. F. Peters. He had just returned from an expedition to observe the transit of Venus when he discovered two asteroids on the night of June 3, 1875. He named them Vibilia, after the Roman patroness of journeys, and Adeona, for the patroness of homecomings.[3]

Christian Heinrich Friedrich Peters was the most successful asteroid hunter of this second phase of discovery. Unlike Goldschmidt, he was a professional astronomer, and he led an adventuresome and contentious life. He was born in 1813 in a small village that was then part of Denmark. He attended the University of Berlin and received his doctorate in 1836. Peters found jobs in astronomy were scarce and finally ended up in Sicily as head of the geodetic survey.

In 1848 the population of Sicily revolted against the rule of the king of Naples. Peters sided with the rebels and served as a major of engineers in the rebel army. The rebellion was crushed by mid-1849, and Peters was forced into political exile, first in France and then in Turkey. When he arrived, Peters had only enough money to buy either breakfast or a cigar—he chose the cigar. He became fluent in Turkish and Arabic and served as scientific adviser to Redshid Pasha. At the suggestion of the American ambassador, Peters emigrated to the United States in 1854 and became an assistant at Dudley Observatory in Albany, New York.

At Dudley Observatory Peters soon found himself caught up in a power struggle between the observatory director, B. A. Gould, and the trustees. Peters backed the trustees and was driven out by Gould in 1858. The trustees found him a job as the professor of astronomy at Hamilton College in Clinton, New York. The move did little to improve his difficult lot. In January 1863 he saw a lawyer about the failure of the school to pay his previous year's salary.

Peters used the 13.5-inch telescope at Hamilton College to observe sunspots by day and hunt for new asteroids by night. His first discovery was made on the night of May 29, 1861 (six weeks after the start of the Civil War). Peters originally thought it was the asteroid Maja (discovered on April 9). T. R. Safford of Harvard was the first to realize it was a new asteroid. He named it Feronia, after the Roman goddess of groves. This began a remarkable series of discoveries by Peters lasting over two decades.[4]

While Peters was the leading discoverer of asteroids, he had a serious competitor. James C. Watson was 25 years younger than Peters. His early years were a time of poverty, which left scars. Watson was both brilliant and ambitious, traits he showed throughout his youth. At one point, his father was an assessor in Canada and became too ill to work. Watson filled in for him, doing all the calculations and drawing up the district's assessment rolls without error. Watson was 9 years old. He subsequently ran the steam engine at the factory where his father worked, sold books and apples at a railroad station (while teaching himself Latin and Greek), and finally enrolled at the University of Michigan at the age of 15.

Watson's first asteroid discovery was made on October 20, 1857, when he independently spotted Aglaja, which had been found several weeks before. After becoming the director of the University of Michigan Observatory in 1863, he prepared a series of star charts covering the ecliptic. Watson also had the ability to memorize the position of every star on a chart.

On September 14, 1863, Watson spotted the asteroid Eurynome. It was another four years before Watson discovered Minerva and Aurora, in August and September 1867. The following year saw his greatest success, with the discovery of six asteroids. For his accomplishments, he was awarded the Lalande Prize of the Paris Academy of Science in 1870.

For the December 9, 1874, transit of Venus, Watson traveled to Peking, China. While there, he made the first discovery of an asteroid from China.[5] Watson gave an account of the event and the court politics that surrounded it:

> On the night of the 10th October, while observing in the constellation Pisces, with the 5-inch equatorial, I came across a star of the 11th magnitude in a region of the heavens with which I was very familiar, and where I had not hitherto seen any such star. Subsequent observations the same night by means of a micrometer, extemporized for the purpose, showed that the star was slowly retrograding, and that it was a new member of the group of planets between Mars and Jupiter.
>
> The discovery was duly announced to astronomers in other lands, and it became also speedily known in Peking. Some mandarins of high rank came to our station to see the stranger, with their own eyes, and upon observing the change of configuration with the neighboring stars on two successive nights, they gave free expression to their astonishment and delight.
>
> This being the first planet discovered in China, I requested Prince Kung, regent of the Empire, to give it a suitable name. In due time, a mandarin of high rank brought to me the document containing the name by which the planet should be known, coupled with a request—communicated verbally—that I would not

> publish the name in China until the astronomical board had communicated to the Emperor an account of the discovery and the name which had been given to the planet. This request was of course promptly acceded to; and I afterwards learned upon inquiry that if the knowledge had come to the Emperor otherwise than through the astronomical board, organized specially for his guidance in celestial matters, some of the ministers would have been disgraced.

The name selected was Juewa, or more fully Jue-wa-sing, which means literally "Star of China's Fortune."[6]

Both Peters and Watson had faced great hardships that produced driven, ambitious, and difficult individuals. The astronomer Simon Newcomb said of Peters, "Of his personality it may be said that it was extremely agreeable so long as no important differences arose." Another astronomer said of Watson, "He was one of the most energetic and able men I ever knew . . . extremely self-confident (but not perhaps more so than his abilities justified), selfish and unscrupulous in advancing his own interests." It was inevitable that they would become rivals.

PETERS, WATSON, AND PLANET VULCAN

Newtonian mechanics was orderly, stable, and rational. Astronomers believed that by using Newton's laws they could account for and understand every movement of the planets. In the early 1850s Leverrier, now the director of the Paris Observatory, embarked upon a mathematical analysis of the orbits of every planet in the solar system. For Leverrier, celestial mechanics and astronomy were one and the same. When his analysis of Mercury's orbit was published on September 12, 1859, there remained a small error. He concluded the cause was an unknown mass, equal to that of Mercury itself, between Mercury and the Sun. Because an object this size could not escape detection, he believed the mass was in the form of numerous small bodies, an inner asteroid belt. The best chance to observe them was when they passed between the Sun and Earth, or during a solar eclipse.

Over the next several years there were sightings of Vulcan, as the planets were soon called, as they crossed the Sun. To the believers, the sightings, along with historical accounts of moving dots on the Sun, seemed to indicate four different Vulcans. Peters was not a believer. At the 1869 solar eclipse, Simon Newcomb, who was going to search for Vulcan, asked Peters to join him. Peters responded that he had come "to observe the eclipse, not to go on a wild goose chase for Leverrier's mythical birds."

While Peters was a complete skeptic about Vulcan, Watson was a believer. Following his final 3 asteroid discoveries in 1877, which brought his total to 22, he turned his attention to Vulcan. The showdown between them came at the Great Solar Eclipse of July 29, 1878. During the eclipse, Watson saw two reddish stars that were not on the star charts he had memorized. They seemed to show disks, and he concluded they were the long-sought Vulcans. Soon after, Lewis Swift, a noted amateur astronomer with several comet discoveries to his credit, announced that he had also seen two unknown planets near the Sun while observing the eclipse from Denver. There seemed to be four Vulcans, as the position of Watson's and Swift's objects were significantly different.

Peters wrote a stinging article on the claims, which was published in early 1879. Peters dismissed the claims made by Swift because he had changed his account several times. As for Watson, Peters argued he had mistaken two stars. Additionally, Peters noted that the objects could not explain the change in Mercury's orbit, as it would take between 1 million and 38 million of them to equal the mass of Mercury. Peters concluded by attacking the reports of objects passing in front of the Sun. These, he argued, were nothing more than short-lived sunspots that observers had mistakenly concluded were moving.

Watson strongly protested Peters's "misstatement of the facts" on May 15, 1879, by saying that his observations were not in error, that Peters had been more than 2,000 miles away, and that he was among the "known enemies" of Leverrier. Watson subsequently dismissed later failures to spot Vulcan with claims he would be vindicated and that Peters would be defeated. It was Peters's arguments that were accepted by most astronomers, however. The numerous failures of transit predictions, along with the poor quality of the sightings and the will-of-the-wisp nature of the whole question, all cast doubt on Vulcan.

Watson himself lived only two years after the controversy began. He died of pneumonia in November 1880; he was only 42 years old. He left the bulk of his $15,000 estate not to his wife, who was left destitute, but to establish a fund to publish tables of the asteroids he had discovered.[7]

Peters outlived his younger rival by a decade. The final two years of his life were spent in a bitter court case over a star catalog prepared by an assistant. Peters won the court ruling. His final asteroid, Nephthys, was discovered on August 25, 1889, a few weeks short of his seventy-sixth birthday. On the morning of July 19, 1890, he was found dead at his doorstep. He had apparently suffered a heart attack as he came home from a night of observing. A burned-out cigar was discovered in his hand.

Although Vulcan now faded into astronomical history, the problem it was meant to solve remained until November 1915. Albert Einstein's gravitational equations of his General Theory of Relativity showed that near a large mass such as the Sun, space was warped, resulting in a change in the orbit of a planet. When Einstein calculated the effect for Mercury, he found it exactly matched the error. Einstein was beside himself with excitement for several days.

Peters's total of 48 discoveries made him the most successful of the asteroid hunters during the second phase of asteroid research. Following behind were Karl T. R. Luther, with 24 between 1852 and 1890, and James C. Watson, who discovered 22 between 1863 and 1877. Peters's final discovery raised the number of asteroids known to 287. This seemingly huge total brought a change in perception about their nature and the practical problems in keeping track of them.

PERCEPTIONS AND NUMBERING SYSTEMS

When the first asteroids were discovered, they were considered to have a status identical with the other planets. This persisted into the second phase of asteroid research. As late as 1857 a reference book stated that there were seventeen planets in the solar system: the eight major planets from Mercury to Neptune and the nine asteroids.[8]

One reflection of this was the procedure of giving a newly discovered asteroid both a name and a symbol, as had been done for the major planets. The symbol for Hebe, the sixth asteroid discovered, was a cup. Although astronomers had long realized that more asteroids would be discovered, they were slow to understand the scale of their numbers. In 1851, following the discovery of three new asteroids, the Council of the Royal Astronomical Society said that so rapid an increase in asteroids "can hardly be expected to continue very long." As more and more asteroids were discovered, the use of symbols was becoming unwieldy. In 1852 Benjamin Valz was the first to use a numbering system when he named 20 Massalia. Rather than a symbol, each asteroid would carry a number and a name.

There was still the problem of asteroids that were observed once but not subsequently sighted again. To deal with these, a new system of provisional designations was established in 1892. New asteroids would be listed by the year and a capital letter (1892 A, 1892 B, 1892 C, etc.). By the following year it was realized that 24 letters were not enough. Accordingly, 1893 Z was followed by 1893 AA. The letters would be assigned according to the date of registration. Later, once an orbit had been calculated, the new asteroid would be given a number and name.[9]

THE THIRD PHASE: PHOTOGRAPHY AND THE HUNT FOR ASTEROIDS

Peters's death in 1890 marked the end of the second phase of asteroid discoveries. The next year, the new technology of photography revolutionized asteroid hunting and began the third phase of asteroid research. The first to apply photography to asteroid hunting was Max Wolf, who began observing from a private observatory at Marzgasse in the old part of Heidelberg, Germany.[10] The power of photography soon became apparent. Wolf made the first photographic discovery of an asteroid on December 22, 1891. It was later given the name 323 Brucia. Over the next year Wolf discovered a total of thirteen new asteroids, one fewer than Goldschmidt had found in nine *years*.

Although the discovery of two asteroids in a single night had been an extraordinary event during the era of visual searches, it became commonplace with photography. On September 25, 1892, Wolf made the first triple discovery when he photographed 339 Dorothea, 340 Eduarda, and 341 California. On January 7, 1896, Wolf found another triple: 411 Xantha, 412 Elisabetha, and 413 Edburga. The high point came on the night of September 7, 1896, when Wolf discovered 418 Alemannia, 419 Aurelia, 420 Bertholda, and 421 Zahringia.

It was also Wolf who photographed 464 Megaira on the night of January 9, 1901, the first asteroid discovered in the twentieth century,

100 years and eight days after 1 Ceres was found. His hundredth discovery, appropriately named 513 Centesima, was found on August 24, 1903. Wolf's lifetime total was 230 asteroids, plus another 20 shared with other astronomers.

The dimensions of this increase in asteroid discoveries become apparent when one looks at the statistics. Between 1851 and 1870 an average of 4 or 5 asteroids were discovered per year. Between 1871 and 1890 this jumped to between about 7 and 12 per year. By 1890 numbered asteroids totaled 302. In the five years from 1891 to 1895, 107 numbered asteroids were found, for an average of just over 21 per year.[11]

In 1896 Wolf was named director of the Konigstuhl Observatory in Heidelberg, a position he held until his death in 1932. Wolf made Heidelberg a center for asteroid research. One of his colleagues was Karl Reinmuth, who joined the observatory staff as a volunteer in 1912. His first asteroid discovery was 796 Sarita on October 15, 1914 (two months after the start of World War I). He remained on the staff until his retirement in 1957. His lifetime total was 389 numbered asteroids.[12]

Others outside Heidelberg were also following Wolf's lead. At Urania Observatory in Berlin, Gustav Witt began a photographic program in 1896. Its goal was not to make new discoveries but rather to photograph known asteroids that had not been observed for some time. In this age of computer-controlled telescopes and automatic guiders, it is hard to recall the difficulties these early efforts faced. Felix Linke, who guided the telescope, later recalled: "Witt had only a rather primitive instrument for this purpose. The clockwork drive worked middling well only for certain areas in the sky, the areas we used to photograph, but even there it was hardly possible to neglect checking on the guide star for more than a few seconds. Its position had to be constantly corrected."

Early in the new century, a new photographic procedure was developed by Joel H. Metcalf. Rather than tracking on stars, so that they were points and the asteroids showed trails, Metcalf followed the asteroids. The asteroids appeared as points, while the stars trailed. This allowed him to spot asteroids significantly fainter than those found with more conventional procedures, since their light was concentrated in a single point rather than smeared out along a line, and it improved the accuracy of the measurements of the asteroid's position.[13]

Unlike Wolf, Witt, and Linke, Metcalf was an amateur astronomer, the most distinguished of his time. He had become interested in astronomy at age fourteen, when he borrowed the book *Other Worlds Than Ours* from his Sunday-school library. He traded a jackknife and several marbles to a schoolmate for a 3-inch lens, then did odd jobs for his mother to collect the money for a mounting.

After becoming a Unitarian minister, he bought a 7-inch Clark refractor at an estate sale for $500 (a major sum for a minister with a wife and two children). Rather than ship the telescope over 100 miles of snow-covered roads around Lake Champlain, Metcalf decided to have it carried across the frozen lake by horse-drawn sleds. The caravan had completed most of the 10-mile journey when the ice gave way beneath the sled carrying the telescope. The box slid off and ended up straddling the crack in the ice. No one dared approach the box lest it slip off and sink into deep water. It remained in this position for nearly a week until the opening froze over. Metcalf took the telescope with him as he and his family moved from parish assignments throughout Vermont and Massachusetts.

Soon after the turn of the century, Metcalf became interested in photographing asteroids. Each fall, winter, and spring, he would use a 12-inch refractor to take photos. Asteroid hunting was a family affair: his son would guide the telescope, while Metcalf searched the plates for new objects. Metcalf discovered 41 numbered asteroids between 1905 and 1914.

While most of the year was given to asteroid hunting, Metcalf spent his summers searching for comets. Metcalf's most famous achievement came in August 1919. The Metcalf family was spending the summer at the Unitarian church camp on Grand Isle in Lake Champlain. Very late on the night of August 20, he spotted comet Brorsen as it returned for the first time since 1847. The object was subsequently renamed comet Brorsen-Metcalf. The following night, August 21, he resumed searching and after only a brief time found another comet. This proved to be the return of comet Kopff (it had been spotted several weeks before). On August 22 Metcalf was again at his telescope. After only a few sweeps, yet another comet appeared, a new comet independently found the next night by Alphonse Borrelly in France. It was Metcalf's sixth and final comet discovery.[14]

One of the most successful asteroid hunters of the late nineteenth and early twentieth century was Auguste Charlois. Born in 1864, Charlois joined the staff of the Nice Observatory in France in 1881 as an assistant.

His first discovery was 267 Tirza, on May 27, 1887. Charlois's first 27 discoveries were made visually, including 297 Caecilia and 298 Baptistina on September 9, 1890. Then, in 1892, a few months after Wolf began his photographic searches, Charlois began a similar program.

Between 1892 and 1898 Charlois and Wolf virtually monopolized asteroid discoveries. Only a handful were found by other individuals. As did Wolf, Charlois made multiple discoveries on a single night; 374 Burgundia, 375 Ursula, and 376 Geometria were all found on September 18, 1893. His final total was 99 numbered asteroids between 1887 and 1904. He never made 100, and in 1910 he was murdered.[15]

While photographic searches quickly became the preferred method of asteroid hunting, one astronomer still used the visual approach, achieving a success that will never be matched. The last visual discoverer of asteroids was Johann Palisa. Born in 1848, he joined the staff of the Pola Naval Observatory, located on the northern Adriatic Sea. At the time, it was the major naval station of the Austro-Hungarian navy. (Since then, Pola became first part of Italy, then Yugoslavia.) His first discovery, made on March 18, 1874, was 136 Austria, the first asteroid discovered in Austria.

In 1881 Palisa moved to the Vienna Observatory, where the chief instrument was a 27-inch refractor, then the largest telescope in the world. He used it for visual searches for asteroids, finding 262 Valda and 263 Dresda on November 3, 1886, and 291 Alice and 292 Ludovica on April 25, 1890. Between 1893 and 1905, when he discovered 569 Misa, Palisa made no discoveries. Successes remained rare until Wolf began supplying him with prints. These made perfect star charts and greatly assisted Palisa in spotting any new asteroids. In October 1911 Palisa discovered five asteroids, three of them (723 Hammonia, 724 Hapag, and 725 Amanda) on the night of October 21. This was the only case of a triple visual discovery.[16] Although Palisa continued searching for the next fourteen years, he faced increasing difficulties. He made between one and four discoveries per year—Wolf often found that many in a single night.

Steady improvements had been made not only in photographic emulsions but also in examining the plates. Wolf's original method was to lay the two plates one on top of the other on a retouching frame, then view them with a magnifying glass. This was unreliable and uncertain. By the early twentieth century, the Blink-Microscope-Comparator had been developed. The two plates were set side by side and the same small area on each was viewed through a microscope. A rotating shutter would cause first one plate, then the other to be seen in rapid succession. The result was that any object that moved between the two exposures seemed to blink as it jumped back and forth. Palisa's classical visual approach could never compete with such industrialized discovery.

The international situation also made his efforts difficult. The Austro-Hungarian Empire had been one of the casualties of World War I. The postwar years brought economic disruption and revolution, while much of its former lands were now split up between the new states of eastern Europe. The difficult situation was reflected in the name of asteroid 964 Subamara (discovered on October 27, 1921), Latin for "very bitter." Palisa's final total was 121 asteroids, every one discovered visually. Since Palisa's death in 1925, no asteroid has been discovered visually.

REVISED NUMBERING SYSTEMS

The flood of asteroid discoveries produced by the third phase required a new numbering system. The old system of using the year of discovery and two letters had a weakness: the two letters did not start over again with the new year. Thus asteroid 1893 AP was followed by 1894 AQ. This caused confusion as to which asteroid was discovered first. It was not until 1924 that a uniform system was developed, which remains in use today. E. C. Bower of the U.S. Naval Observatory proposed dividing the year into 24 half-month intervals, designated by letter, as in table 1.

As each new asteroid was discovered during a particular half-month, it would be given a provisional designation of the year, the half-month, and its order of discovery. Thus the first asteroid discovered between January 1 and 15, 1925, would be 1925 AA, the second would be 1925 AB, and the third 1925 AC. Should 25 asteroids be discovered in any half-month, a number would be added after the two letters—1925 AA1. The system had the advantage of being chronological (even giving an idea of the date of a discovery) and was able to accommodate any number of asteroids. It was also retroactively applied to observations made before 1924 that had not yet been given a designation under the old system. In this case, an "A" would replace the "1" in the year; for example A923RH, which subsequently became 4615 A923 RH. As before, an asteroid would later be given a name and a permanent number once an orbit had been calculated.

Table 1 BOWER'S ASTEROID DISCOVERY DESIGNATIONS

Half-month of Discovery	Letter Designation	Half-month of Discovery	Letter Designation
January 1–15	A	July 1–15	N
January 16–31	B	July 16–31	O
February 1–15	C	August 1–15	P
February 16–29	D	August 16–31	Q
March 1–15	E	September 1–15	R
March 16–31	F	September 16–30	S
April 1–15	G	October 1–15	T
April 16–30	H	October 16–31	U
May 1–15	J	November 1–15	V
May 16–31	K	November 16–30	W
June 1–15	L	December 1–15	X
June 16–30	M	December 16–31	Y

Note: "I" and "Z" are not used.

THE GEOGRAPHY OF THE ASTEROID BELT

Throughout both the second and third phases of asteroid research, astronomers also were developing a better understanding of the asteroids' orbits and the first information about their physical nature. The vast majority of asteroids orbited between Mars and Jupiter—the "main belt" asteroids. Daniel Kirkwood, an astronomer and professor of mathematics at the University of Indiana, realized that the orbits were not random, however. There were gaps in which no asteroids ventured.

He first commented on this at an 1866 meeting of the American Association for the Advancement of Science. Kirkwood published a preliminary paper in 1867 and a more complete account a year later using the orbits of 97 asteroids. He realized the gaps were the result of the gravitational attraction of Jupiter. He wrote: "A planetary particle . . . —in the interval between Thetis and Hestia—would make precisely three revolutions while Jupiter completes one (the 1/3 resonance), coming always into conjunction with that planet in the same parts of its path."

The even ratio causes Jupiter's gravity to nudge the asteroid repeatedly in the same direction, much like an adult pushing a child on a swing. Eventually, the asteroid orbit would be so changed that the ratio there would no longer be even. At this point, Jupiter's gravitational force would be acting at random on the asteroid, and its orbit would be stable.

Since then, a number of Kirkwood gaps, as they have become known, have been found at the 1:3, 2:5, 3:7, 1:2, and 3:5 resonances. Percival Lowell best described the result: "If the asteroids were numerous enough we should actually behold in the sky a replica of Saturn's rings." Three other resonances—3:2, 4:3, and 1:1—have the opposite effect, resulting in a *gathering* of asteroids.

Kirkwood continued to study asteroid orbits, and in 1888 he noted four pairs of asteroids that had nearly identical orbital elements. He quickly realized an insight into the formation of the asteroids and subsequently speculated that these pairs were the result of the breakup of several original parent bodies.

Kirkwood's initial work was extended by W. H. S. Monck of Dublin. He broadened the search, looking for asteroids with similar elements. The result was more than a dozen pairs and an insight into the geography of the asteroid belt. He wrote, "A glance at the list will show that the resemblance frequently extends beyond a single pair and embraces what may be called a family." Although it was Monck who first suggested the existence of asteroid families, his idea attracted no attention.

It was Kiyotsugu Hirayama in 1918 who put the notion of asteroid families on a solid basis. Hirayama looked for families using three of their orbital elements: the semimajor axis (the mean radius of the asteroid's orbit), the eccentricity (how much the orbit differs from a perfect circle), and the inclination (the angle between the orbit and the ecliptic). When Hirayama plotted the results, he initially found three families: Themis (22 members), Eos (21 members), and Koronis (13 members). A subsequent 1923 paper added two more families, Maria and Flora. His analysis seemed to indicate the possibility of several more families, but the similarities were not then precise enough to make an exact determination. As had Kirkwood, Hirayama believed that each family had originated from a single, large asteroid that had broken into many fragments in the distant past.[17]

Just as we speak of Kirkwood gaps, these relationships between asteroids are called Hirayama families. Seventy years later, more than a hundred such families have been found, which include nearly half of the known asteroids. Kirkwood's original four pairs are now considered to be members of four different families.[18]

While analysis of asteroid orbits proved fruitful, attempts to study their physical nature, such as sizes, shapes, rotation rates, and composition, faced limitations. Through even a large telescope asteroids are indistinguishable from dim stars. Despite this, attempts to measure their size began in the early days of asteroid research. Sir William Herschel's attempt to measure the size of the first two asteroids in April 1802 was followed by numerous other attempts during the nineteenth century. The asteroid's angular size was measured, then the diameter was calculated based on the asteroid's distance from Earth. This was an extremely difficult task, as Ceres varies from 0.27 to 0.69 seconds of arc. The results were uncertain; Herschel derived a value of only 259 kilometers for Ceres, but Schroeter calculated it at 2,526 kilometers in diameter, differing by a factor of nearly ten. In 1894 E. E. Barnard used the 36-inch and 40-inch refractors at the Lick and Yerkes observatories to measure the size of the first four asteroids. Barnard, a skilled observer who had begun as an amateur astronomer, derived values that were the best for the next 80 years.

Johann Schroeter in 1810 was the first to attempt to measure the rotation of an asteroid. He found a 27-hour period for Juno. (This was, in fact, four times the actual period.) Success proved evasive, however. A long-term effort to determine asteroid rotations was made by Henry Parkhurst at the Harvard Observatory. This research, undertaken between the 1870s and 1890s, had negative results. Parkhurst, whose day job was as a stenographer in the New York City Superior Court, never found any light variations in any asteroid. The only exception Parkhurst noted was in 40 Harmonia, and this he believed was caused when it passed over a chimney during the three nights of observations.

The problem with such attempts was that they relied on visual estimates of the asteroid's brightness, compared to reference stars. This gave an accuracy of about one-tenth of a magnitude, which was on a par with the amount of variation. Given the small margin for error, it is not surprising that it took 90 years to confirm asteroid rotation. The first asteroid to have its period determined was 433 Eros in 1901. Initially, it was only known that its magnitude varied over the course of several hours. This was later refined to about five hours, then reduced to two and a half hours. Finally it was realized that Eros showed a double period (a peak, followed by a dip, then another peak and dip), totaling five hours. By 1913, when its period was finally determined, another five asteroids had been found to be variable.

A related aspect was the attempt to use the variations in brightness of an asteroid to determine its shape. The first to realize that asteroids might not be round was Olbers. He suggested that reported variations in Vesta might indicate it was angular, with flat sides reflecting the Sun. Research continued, and in 1906 Henry Russell published a theoretical paper on light variations in asteroids. His paper established that although the period of an asteroid's rotation could always be determined, shape and surface markings were a much more difficult matter.

Along with photography, another astronomical breakthrough of the final third of the nineteenth century was spectroscopy. The light from a telescope was passed through a prism, which split the light into a spectrum. Amid the rainbow of colors were black lines where sunlight had been absorbed by a planet's atmosphere or its surface. The first to make spectroscopic studies of asteroids was Hermann Vogel in Germany. On February 3, 1872, he took a spectrum of Vesta. It showed two lines that Vogel believed were due to an atmosphere around Vesta. Similar observations of Flora in October 1873 failed to show any evidence of an atmosphere. In reality, no atmosphere existed around these asteroids. Despite this false start, spectroscopy gave the first indications that asteroids differed from each other in their composition. In 1929, half a century after Vogel's work, Nicholas T. Bobrovnikoff made spectrums of twelve asteroids. He found that Ceres was bluer than Vesta.

It would be a very long time before the potential of this research was fully realized. Part of the problem was that the technology of the late nineteenth and early twentieth century lacked the precision needed. Barnard's measurements of asteroid size, for example, were made visually. Bobrovnikoff's spectrographic study was later described as "so far ahead of its time that it was overlooked."[19]

ASTEROID STUDIES IN DECLINE

Throughout the 1930s the rate of asteroid discoveries continued at a high level. Between 1931 and 1935, an average of just over 43 were discovered per year. The following five years saw a drop to just over 32 per year. On the surface, things seemed to be much as they had been for the previous century. Yet asteroid research, along with planetary astronomy itself, was in deep trouble.

The process had begun 40 years before, at the end of the nineteenth century, with the controversy over the question of life on Mars. In 1895 Percival Lowell's book *Mars* was published. Lowell, a member of one of the wealthiest families in America, had established a private observatory the previous year at Flagstaff, in what was then the Arizona Territory. Lowell's observations led him to conclude that intelligent life existed on Mars. The Martians, Lowell believed, were forced by their dying planet to build a network of canals to transport water from the polar ice caps.

The Mars controversy continued until Lowell's death in 1916. The effect on planetary astronomy would last far longer, however. The response of many astronomers was emotional and vitriolic. The Lowell Observatory became an outcast within the astronomical community, planetary astronomy went into decline, and the amateur's role in astronomy ended.[20]

At the same time, astronomy itself was changing. By the 1930s construction was under way on the 200-inch telescope on Palomar Mountain, the largest, most expensive scientific instrument ever built up to that point. Research into the nature of the atom revealed the process that powered the stars. The attention of both theorists and astronomers now was focused on the grand structure of the universe, and there was little room or interest for the study of mere rocks within our own solar system.[21]

The disdain some astronomers held for asteroids had long been apparent. Metcalf, a tireless discoverer of asteroids, noted in 1912, "The rapid and continuous multiplication of discoveries, since the invention of the photographic method for their detection, has introduced an embarrassment of riches which makes it difficult to decide what to do with them. Formerly the discovery of a new member of the solar system was applauded as a contribution to knowledge. Lately it has been considered almost a crime." A German astronomer best summed up this hostility when he referred to a "plague of minor planets," a phrase that was soon translated as "vermin of the skies."

In September 1939 World War II began, and asteroid research virtually ended for the next three decades. The numbers give the clearest indication of this; by 1976 only 24 asteroids discovered between 1940 and 1945 had been given permanent numbers. None was discovered in 1945, and only 1 each in 1944 and 1946. The number of discoveries would slowly pick up between 1947 and 1950 but remained only slightly greater than during the prephotography days of 1845 to 1890.[22]

3

THE MODERN ERA

By the late 1960s . . . young astronomers wanted to learn about the asteroids (perhaps to the chagrin of their professors). . . . Most of all, they learned that the asteroids were quite exciting!

Charles T. Kowal, 1988

THE END OF WORLD WAR II saw the western half of Germany occupied by the United States, Britain, and France, while the Soviet Union controlled the eastern half. Germany had been a leading center of asteroid research. The Rechen-Institut in Berlin had begun keeping track of asteroids in the 1890s. In 1911 it began publishing predictions of asteroid positions, followed in 1926 by the *RI Circulars,* which carried updated information. The war's end found these activities scattered. The Rechen-Institut had fled the advancing Red Army, moving from Berlin to Heidelberg. About half its staff and material remained in Soviet-controlled areas, however, and it was no longer able to publish. The German observatories that had undertaken asteroid work were also unable to continue, as they lacked even photographic plates. Indeed, for academics and the population at large, life centered on simply finding enough food.[1]

REBUILDING ASTRONOMY IN THE POSTWAR YEARS

The astronomical community faced the challenge of recovering from the wartime disruption. Astronomers were returning from their wartime defense research, students were resuming studies interrupted by the war, and work was being restarted on the 200-inch telescope at Palomar. Those activities formerly undertaken in Germany would have to be continued elsewhere. In early 1946 the International Astronomical Union (IAU) held a meeting in Copenhagen, Denmark, to deal with the postwar situation. At the meeting, the first since 1938, most of the German activities were assigned to Soviet astronomers and observatories.[2]

The exception was asteroid research. In 1947 the IAU established the Minor Planet Center, headed by Paul Herget, at Cincinnati Observatory. It undertook three main activities: publishing *Minor Planet Circulars,* collecting and maintaining asteroid observations, and calculating asteroid orbits and their positions. Herget also had to deal with problems that had been developing for several decades. In 1950 he observed, "The minor planet program was very badly disorganized." While major efforts had been made to discover new asteroids, much less effort had been given to keeping track of the approximately 1,600 numbered asteroids. Some had not been observed for years or even decades. Without updated position data, a large number were considered lost.

Herget's primary efforts were directed at recovering these lost asteroids. In the past an asteroid would be given a name and number as soon as an orbit had been calculated; now the standards were raised by Herget, requiring that an asteroid be observed at two or three oppositions (the point in its orbit when it is nearest Earth). Over the next eighteen years, until 1965, Herget assigned permanent numbers to only 100 new asteroids.[3]

ASTEROID SURVEYS, 1949–1960

In the larger astronomical community, the primary effort in the postwar years was focused, as it had been before the war, on deep-sky observations of galaxies. One individual who kept planetary astronomy alive in these wilderness years was Gerard P. Kuiper. Born in the Netherlands in 1905, he studied astronomy at the University of Leiden in the late 1920s.

He came to the United States and subsequently took up a position at the new McDonald Observatory in Texas. This was an observatory with an eventful history. When William J. McDonald left the bulk of his estate to the University of Texas to build an astronomical observatory, it did not sit well with the other heirs; they contested the will, arguing that funding an observatory was proof of mental illness. Their attempt was unsuccessful. Under Kuiper, McDonald Observatory became a center for planetary astronomy. At one point in the 1960s it was the only observatory to allow planetary observations with large telescopes.

One problem with asteroid studies up to this point was that the numbered asteroids did not give a clear picture of their distribution in space. Bright nearby asteroids were more likely to be discovered than faint ones. From 1950 to 1952 Kuiper directed an asteroid survey at McDonald using a 10-inch telescope that recorded asteroids down to a magnitude of 16.5. The entire ecliptic was photographed twice. Once corrections had been made for incompleteness, the photos provided statistical data on asteroid populations. This initial survey was followed by detailed measurements of all asteroids with a brightness greater than about tenth magnitude. Two other asteroid surveys were undertaken at this time. The first was the Indiana Asteroid Program, started in 1949 by F. K. Edmondson; lasting a decade, it provided photographic magnitudes of asteroids. The other was the University of Finland survey begun by Yrjo Vaisala in 1935. It was virtually the only asteroid survey conducted during the war, and it continued until 1957.

Their efforts were followed by the Palomar-Leiden Survey in 1960, which used an approach different from the earlier efforts. Rather than surveying the entire sky, a smaller area was observed to a much fainter magnitude. The result was brightness and distance measurements of some 1,800 asteroids. The asteroids observed during this survey were given a special designation: a four-digit number followed by the abbreviation "P-L."

TOM GEHRELS AND THE REBIRTH OF ASTEROID STUDIES

Tom Gehrels was one of the astronomers involved with the Palomar-Leiden Survey. More than any other individual, Gehrels was responsible for the rebirth of asteroid studies in the postwar era. Like Kuiper, he was born in the Netherlands. Only fifteen years old when Nazi Germany invaded Holland in 1940, he escaped to England, then returned by parachute to join the Underground. After the war, he developed an interest in astronomy and became a student at Leiden. In 1950 Gehrels came to the United States and eventually met Kuiper, who gave him the task of measuring asteroid brightness.[4]

In 1971 Gehrels organized the first asteroid conference at Tucson, Arizona. There were about 140 participants at the five-day meeting—astronomers, geologists, chemists, engineers, and physicists specializing in asteroid studies. The conference came at a time when asteroid studies were ready for a rebirth. The problems of the immediate postwar years were largely solved, the U.S. space program had opened new research possibilities, and students were attracted by the new and wide-open field of planetary science. Although some older astronomy professors still advised their students to avoid the field, the professional roadblocks that had discouraged such work were gone.

New tools were also now available to asteroid studies. In the nineteenth century observers had only their eyes to measure an asteroid's brightness. This was accurate to about a tenth of a magnitude. Photography allowed more-precise measurements to be made. The iris photometer measured the brightness of stars in photographs based on how much light came through the image of a star on a glass negative. This was accurate to about five hundredths of a magnitude but required a long exposure.

It was the electronics revolution that opened new horizons in astronomy. The first breakthrough was the development of the 1P21 photomultiplier tube by RCA during World War II. The faint light of an astronomical body would be multiplied several times within the tube, then transformed into an electrical current. The 1P21 photomultiplier tube was first used in astronomy in the early 1950s by W. W. Morgan and Harold Johnson. The procedure was called differential photometry. Three measurements were taken of a star through ultraviolet (U), blue (B), and visual (V) filters. In addition to the star under study, a comparison star of a known brightness was measured, as well as a nearby area of empty sky. The "integration time" of each observation was only a few seconds, rather than the minutes or hours required for a photographic plate. The UBV system could provide an accuracy of 0.001 magnitude, and it quickly became the standard method for measuring star brightness.

The second great electronic breakthrough was the use of computers to process data. When the UBV system was introduced, each of the hundreds of observations from each night had to be manually corrected for the air mass (the thickness of air the light had to pass through), while the background sky brightness had to be subtracted. In the late 1960s and early 1970s, mainframe computers, which could do these calculations faster and more accurately, began to appear at universities. The data from the telescope would be printed out on a teletype and punched into a paper tape, which would then be used to make punch cards. A night's observations would produce a stack of punch cards four or five inches high to be dropped off at the university's computer center for processing. After the astronomer had made several visits to the computer center, the light curves would be printed out and ready for analysis.[5] Astronomers now had the means to process huge amounts of very precise data into a usable form. The way was clear for a revolution in our understanding of asteroids.

THE PHYSICAL PROPERTIES OF ASTEROIDS

The most basic physical property of the asteroids that could be measured with the new tools was their rotation rate. Attempts to detect light variations in asteroids date from the early years of research. In 1810 Schroeter thought he detected a 27-hour rotation in 3 Juno (about four times the current value of 7.21 hours). Henry Parkhurst spent three decades observing asteroids but was unsuccessful in detecting any light variations. It was not until 1901 that E. von Oppolzer was able to confirm brightness variations in 433 Eros.

The problem was that light variations were small and difficult to detect except under ideal conditions. In some cases, the amplitude of the variation is as small as a few hundredths of a magnitude. Most range from 0.2 to 0.4 magnitude. The largest recorded variation is 2.03 magnitudes, by 1620 Geographos. The shapes of the individual light curves also show wide variations. Some are simple sine waves; others show a sudden spike in brightness, followed by a steep drop-off to a lower level, which then remains fairly constant.[6]

To complicate matters further, the same asteroid may show completely different light curves at different oppositions. This is because of the different and changing geometric relationship between the Sun, Earth, asteroid, and the asteroid's spin axis. To take two extreme cases, if the asteroid's spin axis was pointed directly at Earth, there would be no change in the light curve, but if it was at a right angle to Earth, the variations would be the maximum value. An asteroid has to be observed over several oppositions to determine the orientation of the spin axis. Even with this corrected for, finding the rotation period can still be difficult. An asteroid may have one, two, or even three brightness peaks during each rotation.[7]

Most asteroids were found to have rotation periods of between 6 and 13 hours. For many years the shortest value known was 2 hours and 27 minutes for 1566 Icarus; it was not surpassed until the discovery of 1995 HM, which had a period of under 2 hours. That record was shattered by 1998 KY26, a tiny spherical asteroid only 100 feet in diameter; its day is only about 10.7 minutes long, or 135 times faster than Earth's rotation, resulting in a sunrise or sunset every 5 minutes. Astronomers believe that this very high rotation rate was due to collisions with other asteroids.[8]

As of the late 1980s only fifteen asteroids were known to have periods greater than 40 hours. The record holder is 288 Glauke, with a period of 47 days. Alan Harris and James Young of the Jet Propulsion Laboratory (JPL), who led a program of observations, noted that the number of slowly rotating asteroids was greater than could be expected by chance. In some cases, the slow rotations could be explained by outgassing, or some frictional process, such as tidal effects from a moon. For 288 Glauke, however, Harris said that it "is clearly a remarkable anomaly, as yet without satisfactory explanation." He did offer a whimsical explanation: "The inhabitants of Glauke became tired of Nature's uniform gray and chose to repaint their planet, first with a dark primer during April, and then with a bright new color, which they completed around mid-May. Later that month, the paint began peeling off to reveal the old gray again."[9]

An asteroid's light curve can also be used to determine the object's shape. The first approach is to match its light curve with that of a model; the other is to derive a shape from the light curve itself. In 1971 J. L. Dunlap made the first laboratory tests of the specific light curves that different shapes would produce. Twelve different models were built, using various cylindrical shapes with hemispherical ends. They had a Styrofoam body, covered with a thin layer of plasticene and then sprinkled with powdered rock. To simulate areas of reduced reflectivity (called albedo), the models were covered with dark graphite powder. The models were illuminated, then turned with a motor. The reflected light was measured with a photometer, like that used in a telescope.

A decade later, two Italian astronomers, M. Antonietta Barucci and Marcello Fulchignoni, extended this work. Their study was called System for Asteroid Models, or SAM. Their models could vary the aspect, obliquity, and phase angle, which better simulated the different observing conditions. This required literally thousands of light curve measurements, however.

One surprising finding was that a large crater on an asteroid did not change the light curve as long as its albedo was the same as the surrounding area, which would occur if the asteroid was uniform in composition. If, however, the interior had a different composition from the surface layers, the crater would be apparent in the light curve. One of the SAM light curves was rather unusual. Barucci had Fulchignoni stand next to the model support, then turn around. The result was the first human-head light curve. Barucci said of the results, "The unusual asteroid (half very dark and the other half with mountains and craters) showed the nicest light curve that we know from the literature!"

Another procedure was to derive the shape directly from the light curve, a method used by Steven Ostro and Robert Connelly in 1984. Called convex-profile inversion, it produces a two-dimensional convex profile, which can then, under certain conditions, be used to derive a three-dimensional shape.[10]

The ultimate test would be to resolve the shape of an asteroid. In the mid-1970s a new procedure, called speckle interferometry, was developed to make high-resolution images using ground-based telescopes. The large telescopes must make exposures lasting only a few thousandths of a second, thereby freezing atmospheric effects that would otherwise blur the shape of the asteroid. Each frame records a distribution of speckles of tiny images. Frames taken over several minutes are combined into a single image by a computer.

In 1981 astronomers at Steward Observatory in Arizona began a speckle interferometry program to check estimates of asteroid size, shape, and materials derived from indirect methods. The initial images of 433 Eros showed only blobs of light. After they were processed, the asteroid showed an oblong shape. In a series of images, the asteroid could be seen to rotate.[11]

Based on observational data and theoretical calculations, the shapes of a number of asteroids had been determined by the early 1990s. Both 1 Ceres and 2 Pallas were nearly spherical. Other spherical asteroids included 6 Hebe and 8 Flora. Most asteroids were found to be elliptical in shape, however, including the other two original asteroids, 3 Juno and 4 Vesta. The asteroid 39 Laetitia was a triaxial ellipsoid 255 × 150 × 85 kilometers; 1580 Betulia was a prolate spheroid (i.e., football-shaped). Finally, there are a few asteroids whose shapes are, well, odd. The asteroid 93 Minerva, for example, is described as pancake-shaped.

THE ASTEROID ALPHABET SOUP

As much as can be learned from measuring the brightness of the light reflected from asteroids, it is small compared with what can be discovered from *how* the light is reflected. The different wavelengths of sunlight illuminating an asteroid will be either reflected or absorbed, depending on the specific material it is made of. Those reflective properties can then be compared to those of different types of meteors and minerals, making it possible to determine the asteroid's composition.

It was in the early 1970s that Thomas McCord, Clark Chapman, and a few others made the first detailed studies. The asteroids were observed both with UBV filters and in the far infrared. The bulk of the asteroids observed fell into three major classifications. Most common are the C-type asteroids, which resemble the carbon-rich chondritic meteors (thus the "C"). The second group, the S-type asteroids, showed the presence of metal-rich silicates, such as those in stony or stony-iron meteors. The C-type makes up about 75 percent of the asteroids, while the S-type amounts to about 15 percent of the total. The third, much smaller group is the M-type (for metallic). M-type asteroids are thought to resemble nickel-iron meteors.

By the late 1970s the classification system had been broadened with the addition of three more groups. The E-type (for enstatite) has very high albedos (greater than 23 percent of the light striking these asteroids is reflected; only 0.65 percent is reflected for C-type and 7 to 23 percent for S- and M-types). The R-type asteroids are very red objects, with albedos on a par with the E-type asteroids. U-type asteroids are those considered unclassifiable in this system. For example, they show a wide range of albedos. Although they each may be a unique object, the U-type classification itself was ever-popular.

A decade after this effort began, the techniques of spectrophotometry had advanced considerably. Three different groups—Edward Tedesco, David Tholen, and Ben Zellner of the University of Arizona; Tedesco and Jonathan Gradie of Cornell University; and Glenn Veeder, Dennis Matson, and Charles T. Kowal of JPL-Caltech—had all made extensive observations of hundreds of asteroids, at a wide range of wavelengths.

These additional spectral and infrared data caused a revision to the list of asteroid types. Gradie and Tedesco added three new types, and the Veeder group eliminated one type and added another. The new categories proposed by Gradie and Tedesco were the P-, F-, and D-types. The P-type asteroids were very dim objects (albedos of 0.65 percent) but with spectrums similar to M-type asteroids, thus the designation "P," for pseudo-M. The F-type showed flat spectrums with no ultraviolet absorption features, and the D-type asteroids were very red objects in the far infrared. Like the P-type, both these categories showed very low albedos. Veeder's infrared observations suggested that the R-type asteroids did not exist as a separate class. They were actually S-types, except for a very small group of very red asteroids, which were designated A-type. (There were only three identified as of 1983.) They had albedos of 17 to 25 percent and showed an abrupt brightening in near-infrared wavelengths.[12]

In 1985 the C-type was reorganized into four subgroups. In addition to the original C-type, there was also the F-type, the G-type (with strong ultraviolet features), and the B-type (similar to classical C-type asteroids but with albedos twice as high).[13]

Finally, in the late 1980s and early 1990s, three more rare asteroid types were added: the Q-type, which consists of three asteroids that pass close to Earth, the V-type, and the T-type. The complete asteroid alphabet, in descending order of abundance, is C, S, M, D, F, P, G, E, B, T, A, V, Q, and R. To complicate matters further, many asteroids show features of several different types, in which cases a string of letters is used. These include a CU-type, a CMEU, a CPF, and an MP.

Even with all the data, there were uncertainties. One longtime controversy was over the S-type asteroids. The conventional wisdom was that they were the source of stony-iron meteors. During the 1980s, however, a number of astronomers argued that the spectrums of these asteroids were inconsistent with such metal-rich materials. Rather than stony-iron meteors, it was argued that they were ordinary chondrites. It was suggested that some type of space weathering, such as solar radiation or microscopic impacts, could have modified the surface layer and changed its spectral characteristics.

THE ORIGIN OF THE ASTEROIDS

By the late 1980s and early 1990s, the photomultiplier tubes that had revolutionized astronomy were being replaced by the charged coupled device (CCD), an electronic chip with an array of thousands of light detectors. Far more sensitive than a photometer, each of the light detectors can measure a far smaller segment of an object's spectrum. This allowed an understanding not only of the physical properties of asteroids but also of their origins.

The first indication of this came when the distribution of the different types of asteroids was plotted. They did not orbit at random, but within specific parts of the asteroid belt. The S-, M-, and E-type asteroids predominate in the inner part of the belt. The center section was home to the F-, G-, B-, and T-type asteroids. In the far reaches of the asteroid belt are found the C-, D-, and P-type asteroids.[14] Gradie and Tedesco noted the significance of this pattern. They wrote, "If the observed differences among the asteroid types are truly compositional in nature, then the type distribution can be interpreted as representing gross compositional changes across the belt."

The distribution of asteroid types matched that expected from several models of how the solar system condensed out of the original cloud of dust and gas. The silicate- and metal-rich types formed closer to the Sun, while those asteroid types composed primarily of lighter elements, such as carbon, condensed farther out. The same pattern is seen in the planets. Mercury, Venus, Earth, and Mars are all rich in silicates and metals, while the outer planets (Jupiter, Saturn, Uranus, and Neptune) are primarily hydrogen and carbon, with only limited percentages of silicates and metals. The asteroid belt is a dividing area between the realms of the two types of planets, and its composition reflects that change.[15]

The observations made of 4 Vesta over the past three decades provide an example of how these observation techniques have expanded our understanding of the processes that formed and shaped the asteroids. In 1967 Tom Gehrels determined Vesta had a period of 5 hours and 20 minutes. Then, in 1974, R. C. Taylor measured the rotation rate at 10 hours 40 minutes and 58.84 seconds. In 1983 Michael Gaffe, an astronomer and geologist at the University of Hawaii, made the first map of Vesta. He measured the small spectral changes as the asteroid rotated. From these subtle changes, amounting to only a few percent, he was able to determine the material at specific areas on its surface. Gaffe found that most of Vesta's surface was similar to the eucrite meteorites—a calcium-aluminum silicate. He also found two equatorial spots that resembled diogenite meteorites, which are magnesium-iron silicates. One of these spots shows a ring of diogenite with a center of olivine-rich material.[16]

It was not until 1987 that speckle interferometer observations of Vesta made by Jack Drummond, Andreas Eckart, and Keith Hege of the University of Arizona finally cleared up the confusion surrounding Vesta's rotation period. The asteroid was found to be elliptical, with dimensions of 580 × 530 × 470 kilometers. This shape would normally produce two brightness peaks per rotation, supporting the 10 hour 40 minute period, which Gaffe had used to produce his map. But the speckle interferometer data also showed that Vesta had several dark areas on its largest side. Vesta is actually dimmest when its maximum cross section faces Earth. Thus it showed only one brightness peak per rotation and a 5.3-hour day. The light curve of Vesta was controlled by the surface features, not its shape.[17]

The visible light and infrared data from Vesta were a near-perfect match for the eucrite, diogenite, and howardite classes of basaltic achondrite meteorites (which account for about 6 percent of the known meteorites). The eucrites come from the surface layer of Vesta; the diogenite meteorites originate from a layer below the surface. The howardite meteorites are a mixture of the two other types, formed by impacts. Vesta was subsequently placed in a new classification, the first V-type object. A family of nearly two dozen 5- to 10-kilometer asteroids have orbits similar to Vesta's and show the same spectral signature.[18]

Eucrites are basaltic flows, like the thick layers of lava covering much of the Moon's surface, and diogenite is a plutonic rock, formed when molten rock cools slowly deep underground. Their presence indicates that Vesta was melted early in its history.[19] The problem is that Vesta is too small to have generated sufficient heat to produce such rocks by the classic processes of radioactive decay of uranium, potassium, and thorium and gravitational collapse.

The current theory is that the melting was caused by electrical heating. When the Sun was newly formed, it underwent a brief period of increased luminosity lasting between 10,000 and 1 million years. The newly formed Sun was rotating rapidly and giving off a very strong solar wind of electrically charged plasma. The electric and magnetic fields produced by this solar wind created electrical currents within the early asteroids. The strength of the electrical currents in a specific asteroid, and thus the amount of heat, would depend on its size, distance from the Sun, and individual electrical conductivity. The energy would be deposited deep within the asteroid, and over time, enough heat would build up to melt the core. A 200- to 500-kilometer asteroid would require 10 million years for its core to melt. The surface crust would act as an insulating blanket, keeping the core hot for a million years.[20]

Observations of asteroids, study of meteorites, space probes to the Moon and planets, and theoretical calculations have given astronomers a basic idea of the origin of the asteroids. About 4.5 billion years ago the solar system was a huge disk of dust and gas. At its center, the Sun had formed, having completed its collapse into a protostar, and had begun burning hydrogen gas in a fusion reaction. Within the dust cloud of the solar nebula, materials with high melting points, such as iron, condensed into dust first, followed by those with lower melting points, like carbon. In the inner part of what would become the asteroid belt, temperatures remained too high for these materials to condense, and they chemically combined with the higher-temperature elements. In the middle and outer parts of the belt, the low-melting-point carbonaceous materials, as well as ice, were able to condense.[21]

As the dust grains orbited the Sun, they were moving slowly relative to each other. As a result, they would collide with each other and stick together. Once a loose collection of dust grains had formed, these "planetesimals" would become quite efficient at trapping more and more of the surrounding dust grains. Traces of this process have been found in the microscopic structure of meteorites. It is also suggested that in some areas of the solar nebula the dust grains would be close enough to be pulled together into flying dustballs. Depending on the amount of dust, planetesimals as large as several kilometers could be formed.

The planetesimals were ideal for sweeping up material. Like terrestrial dustballs, the planetesimals had a loose, granular structure. The energy of any object hitting them would be absorbed, preventing it from rebounding. The object would be captured within the dustball. Once a few of the planetesimals had become larger than the others, they would begin to grow at an accelerated rate. Their gravitational attraction would begin to sweep up the surrounding dust, and the added mass would extend their reach.

The process would affect objects of all sizes—dust grains would be trapped by tennis-ball-sized objects, which would then be captured by kilometer-sized planetesimals, which in turn would collide with objects made up of several planetesimals. The result would be large planetesimals that were loose aggregates of several smaller objects, all with differing compositions. These bodies could have had large voids between the objects, reducing the planetesimal's density.[22]

As the planetesimals were forming, they were also being heated. Some of the early asteroids formed iron cores, with outer layers of stone, like miniature versions of Earth. In other cases, pockets of iron may have melted within an asteroid but then failed to form a central core; the iron masses remained scattered throughout the asteroid. The rate at which the iron cooled can be determined by analysis of meteorites, which have shown a wide range, from 1 to 10 degrees per million years up to 15 to 250 degrees per million years. As a general rule, the deeper the molten iron was within an asteroid, the slower the cooling would have been.[23]

The asteroids in the middle part of the belt showed lesser degrees of heating and underwent metamorphosis. In the process of being heated, they did not melt but rather lost much of their volatile lighter elements, as well as most of their water. The ice melted and the water circulated through the asteroid before it reached the surface through cracks and was vented into space. For a brief time, the asteroid was transformed into a comet.[24] In the process, silicate grains and glass were transformed into water-rich claylike particles.[25]

From spectral analysis of main- and outer-belt asteroids, it was clear that this production of "aqueously altered silicates" was not a uniform process. Observations of sixteen main- and outer-belt asteroids showed that the farther from the Sun the asteroid is, the fewer the signs of water modification. Another ten asteroids beyond the main belt showed no signs of the process at all. These C-type asteroids, in the dark outer reaches of the belt, were viewed as primitive bodies. They had not changed since they formed, retaining their high carbon and water content.

This early asteroid belt was far different from the one we know today. According to an estimate by David Hughes of the University of Sheffield, the biggest asteroids could have been slightly larger than Mars. He further calculated that the original asteroid belt had a mass 2,200 times greater than that of the current belt. Most of this mass was in a few large objects, rather than in the many tiny bodies we see today. He estimated that there were once at least 600 asteroids larger than 1 Ceres.[26]

When the early asteroids were fully formed, farther out the gas and dust continued to form planetesimals. These underwent runaway growth, continuing to pull in gas, dust, and ice. The gravity of these massive planetesimals, which eventually became Jupiter, had a profound and destructive effect on the ancient asteroid belt. The asteroids were stirred up; their orbits became more elliptical, and the inclinations of their orbits increased. The current orbit of 2 Pallas is inclined nearly 35 degrees. An object as large as this asteroid could not have formed in so highly inclined an orbit. Rather, it was perturbed into this orbit by the close approach of a massive planetesimal.[27]

Due to the disruption of their orbits, the average speed between the asteroids was now 5 kilometers per second. No longer could the asteroids coalesce into larger and larger objects. When asteroids collided at this speed, they did not join together but were shattered into fragments by the impact. As the large bodies were broken up by repeated impacts, the asteroid families were created.

Studies of the individual members of a family can give clues to the original parent asteroid. The Nysa family consists of two large asteroids, 44 Nysa and 135 Hertha, each of which is about 70 kilometers in diameter, and a number of smaller asteroids, each about 20 kilometers in size. The asteroid 135 Hertha is an M-type object, composed entirely of metals such as iron. In contrast, 44 Nysa and the smaller objects are E-type asteroids, composed of enstatite chondrites. It is assumed that 135 Hertha is the remnant of an iron core, and 44 Nysa and the other objects were originally the stony outer crust, reduced to small fragments by repeated impacts.

The Nysa family shows differences between its members, but other families have uniform compositions. The Themis family is all C-type asteroids in the outer belt. The original body is estimated to have been about 300 kilometers in diameter, with nearly the same composition throughout. Similarly, the Koronis family members are all S-type asteroids, originally either from a much larger S-type asteroid that was uniform throughout or from a core object whose mantel had been removed before being shattered.

Examination of meteorites shows evidence of these impacts. Their crystal structure has been deformed by an impact, while some materials have been partially melted. The result is a breccia, a rock composed of broken rock fragments that have been compacted together within a finer grain material. The individual rock fragments that made up the breccia came from all parts of the original asteroid, showing the differences in their original formation, internal structure, and composition.[28]

The original large asteroids were steadily reduced in size by the repeated impacts. The resulting debris was also being cleared out by Jupiter's gravitational effects. Some of the fragments knocked off the asteroids passed within 200 million kilometers of Jupiter and either collided with the planet or were thrown out of the solar system. A few were even captured by the outer planets, becoming moons.

According to calculations made by Chyi Liou of the Johnson Space Center and Renu Malhotra of the Lunar and Planetary Institute, as the asteroids were cleared out, the orbits of the outer planets began to shift. Saturn, Uranus, and Neptune moved farther from the Sun, while Jupiter moved inward. Asteroids that had been in safe orbits were now in resonance with Jupiter. They were pushed into unstable orbits and were cast out of the asteroid belt.[29]

Eventually, as most of the asteroids were cleaned out, the rate of collision began to drop. The remaining bodies were no longer being broken up into smaller fragments. With fewer fragments sent into unstable orbits, the structure and mass of the asteroid belt stabilized. Of the original large bodies, only Ceres, Pallas, Juno, and Vesta survived. The remainder exist today only as fragments. The remaining mass of all asteroids is equal to only one-tenth that of Earth's Moon. Of this, the four large asteroids account for a full 40 percent of the mass of all asteroids a kilometer or larger in size.[30] The asteroid belt we see today is but a ghost of the original. It is the scattered debris left from the destruction of very long ago.

THE LOST ASTEROIDS

While new technology was allowing an understanding of the asteroids' past, work was also under way to track down those that had been lost over the years. The problem had been developing since the use of photography had accelerated the discovery rate. In the late nineteenth century the standards for granting numbers to new asteroids had been casual. It was enough that an orbit was calculated. Follow-up observations were spotty. A number of asteroids were seen only at their discovery opposition and not recovered on subsequent occasions.

In 1903 the scale of the problem was clear to Edward Pickering. He wrote, "Of the five hundred asteroids so far discovered, sixty-eight have not been seen during the last five years, while the last observation of twenty-five of them was from ten to thirty-five years ago. Evidently there is a great danger that many of them will be lost, and then it will be impossible to decide when one is observed whether it is new or not. Finding missing asteroids is evidently much more important than discovering new ones."[31]

By the end of World War II the problem of lost asteroids was too great to be ignored any longer. When the Minor Planet Center was established in 1947, one of Paul Herget's primary activities was to recover them. As a result of his efforts, the number of lost asteroids was cut sharply. In 1979 there were 20 lost asteroids, and only 11 in 1981. They were 330 Adalberta, 452 Hamiltonia, 473 Nolli, 719 Albert, 724 Hapag, 843 Nicolaia, 878 Mildred, 1009 Sirene, 1026 Ingrid, 1179 Mally, and 1537 Transylvania. In each of the 11 cases, the asteroids had been seen only at their discovery opposition.

Three of the missing asteroids were discovered in short order. On February 15, 1981, asteroid 1537 Transylvania was recovered in a photo taken with the 48-inch Schmidt telescope at Palomar Observatory. It had originally been discovered at the Budapest Observatory and had not been seen again for nearly 41 years. Later that year, 452 Hamiltonia was recovered after being missing for twice as long. It had been discovered by James Keeler at Lick Observatory on December 6, 1899, and had not been seen since. The two missing asteroids' positions were calculated by L. K. Kristensen of the University of Aarhus in Denmark.[32] At the end of May 1981 the asteroid 843 Nicolaia was recovered by H. E. Schuster at the European Southern Observatory. Nicolaia was found only 75 arc minutes from the location calculated by Lutz D. Schmadel.

Early in 1982 a fourth lost asteroid was removed from the list. The asteroid 330 Adalberta had earned a difficult reputation. The only observations were on the two discovery photographs, taken on March 18 and 23, 1892, by Max Wolf. It was given the initial designation 1892 X. With just the two positions, only a rough circular orbit could be calculated. Despite this, and the lack of additional observations, the asteroid was later given the permanent designation 330 Adalberta.

Adalberta then embarked on its 90-year career as a nightmare for orbital calculators. In 1932 R. Hiller proposed that Adalberta and a newly discovered asteroid, designated 1932 DB, were the same object. A few years later, Gustav Stracke was able to show, despite the rough nature of Adalberta's orbit, that it could not be 1932 DB. Five decades later Lutz D. Schmadel, of the Astronomisches Rechen-Institut, and Richard M. West and Claus Madsen of the European Southern Observatory attempted to match the position of Adalberta with the March 1892 locations of some 2,500 numbered asteroids. Again, there was no match.

Finally, Schmadel, West, and Madsen decided to reexamine the original photographic plates, which were now at Konigstuhl Observatory in Heidelberg. The two plates had survived two world wars and the passage of 90 years. They also found a third plate of the same area. They then located the two asteroid images and compared them to modern photographs. They discovered that the supposed asteroid on each of the two plates was two different stars. This created the illusion of a moving object. Adalberta never existed!

The mistake had occurred for several reasons. The two plates had different emulsions, which means that they had different sensitivities to dim stars. One of the stars was not recorded on two of the three plates. Second, Wolf was overloaded with work, trying to photograph asteroids by night, then examining the plates by day. He had simply mistaken the two stars for a single moving object and reported it as an asteroid.

Another reason was the particular situation surrounding the number 330. It had originally been given to an asteroid discovered by Max Wolf on March 19, 1892, which was named Ilmatar. It was soon realized that this new discovery was actually asteroid 298 Baptistina. The name "Ilmatar" was switched to 1894 AX, which Wolf had discovered on

March 1, 1894, and was given number 385. Number 330 was now empty, and to fill the gap, it was switched to 1892 X, despite the poor quality of its orbit. To replace the nonexistent discovery, another asteroid discovered by Wolf, A910 CB, was given the ill-fated designation 330 Adalberta.[33]

With 330 Adalberta eliminated, the number of missing asteroids was now reduced to seven. Only a few more months would pass before another was recovered. On July 14 and 15, 1982, asteroid 1009 Sirene was photographed using the 48-inch Schmidt telescope on Palomar Mountain. The recovery was made possible by calculations by L. K. Kristensen. The number of missing asteroids had been cut from eleven to only six in a year and a half.[34] It was to be another three and a half years, however, before the list would again be reduced. Then, in December 1986 and January 1987, three missing asteroids were announced as recovered.

The three were 473 Nolli, discovered by Max Wolf in February 1901; 1026 Ingrid, found by Karl Reinmuth in August 1923; and 1179 Mally, originally observed by Wolf in March 1931. Mally was recovered after considerable work by Schmadel and West. They began by remeasuring the plates and calculating a new orbit. Despite this, Mally could be anywhere within a large area of sky. In March 1986 they asked Schuster to photograph the predicted area with the 1-meter Schmidt telescope at the European Southern Observatory. The photographs showed hundreds of asteroid trails, and Schmadel and West had to sort through each one to find any that might be Mally. When a possible object was found, Schmadel was able to connect it with the 1931 discovery observations, while West found images from 1952, 1979, and 1983.

The next to be recovered was 1026 Ingrid. There were only four positions from the 1923 discovery. It was a Japanese amateur, Syuichi Nakano, working at the Smithsonian Center for Astrophysics, who was able to link the 1923 positions with two asteroid trails accidentally recorded by Palomar Observatory's 48-inch Schmidt on March 6 and 8, 1986. The identification was confirmed when two more single images from 1957 and 1981 fit the new orbit.

The third of the lost asteroids to be found was 473 Nolli, which had been missing since 1901. Late on New Year's Eve 1986, Minor Planet Center director Brian G. Marsden was finally able to identify Nolli with asteroids observed in 1940, 1981, and August 1986. This was possible because West had remeasured the original 1901 discovery plates. Marsden found that Nolli's orbit was considerably smaller than had previously been calculated and that it was also about two magnitudes fainter than expected.[35]

This flurry of recoveries had again cut the number of lost asteroids in half, from six to three. The remaining lost objects were 719 Albert, 724 Hapag, and 878 Mildred. It was nearly two years before the next was found. Syuichi Nakano was able to show that an object observed on November 8, 1988, by Tsutomu Hioki and Nobuhiro Kawasato was 724 Hapag. This was the first known sighting since the asteroid's discovery by Johann Palisa in 1911.[36]

The final recovery of a lost asteroid was that of 878 Mildred. It had been discovered on September 6, 1916, by Seth B. Nicholson and Harlow Shapley, at the Mount Wilson Observatory. Shapley named the asteroid for his infant daughter, Mildred. The asteroid was dim and could be observed for only six weeks. The resulting observations were not sufficient to calculate a reliable orbit, and it was lost for the next 75 years. Finally, Gareth V. Williams of the Minor Planet Center identified a dim trail on an April 10, 1991, photo as Mildred, which allowed him to locate other images. The first was from 1985, then a second image from 1984 was found, and finally a third from 1977, which clinched the identification. Marsden said later, "This was a very clever piece of computational detective work by Gareth, from truly minimal information."

The recovery was also a relief to Mildred Shapley Matthews, the asteroid's namesake, who was working at the Lunar and Planetary Laboratory as an editor of the University of Arizona's Space Science Series. She was delighted to learn that she would no longer have to put up with comments that "Mildred is lost."[37]

There remains only one missing asteroid, 719 Albert. The discovery was made on October 3, 1911, by Johann Palisa, and was given the initial designation 1911 MT. Palisa discovered five asteroids that month, but 1911 MT stood out because of its fast motion of 0.75 degrees per day, indicating that it was passing relatively close to Earth. The problem started immediately. Palisa spotted a pair of dim stars and realized that one of them was an asteroid. As he attempted to measure their position, clouds moved in, and Palisa had to wait half an hour for the sky to clear. When it did, the southern of the pair had noticeably shifted. Palisa had less than half an hour to observe the new asteroid before the sky clouded up for the rest of the night.

The next evening, under bright moonlight, Palisa again observed 1911 MT. Three hours later, as the Moon was setting, C. H. Pechule at the Copenhagen Observatory made two measurements of 1911 MT's position with the 14-inch telescope. He said, "It was very faint in a foggy sky. Between the two observations, it passed so close to the west of a star of similar brightness that I could no longer distinguish it clearly. After it had moved far enough south of this star, I obtained the second observation, which was less certain than the first, because the planet had become very faint at an altitude of 20 degrees." The next several nights were plagued with cloudy skies and increasing moonlight. After the full moon had passed on October 8, Palisa again searched for 1911 MT but never saw it again. During October, several observatories also made photographic searches for the asteroid, but all were failures.

Only four positions had been measured for 1911 MT: one on October 3 and three more on October 4 that covered a span of four and a half hours. A rough orbit was calculated, which allowed 1911 MT's positions during the previous summer and fall to be estimated. Astronomers went back to their archives, looking for any photos that might show the asteroid. The initial results looked promising. Charles R. Davidson at the Royal Greenwich Observatory found three images of 1911 MT on plates taken with the 30-inch telescope on October 11. At Heidelberg, a pair of plates that had been taken with the 16-inch double refractor on October 17 apparently showed 1911 MT. The images were very weak, however, and the photos were considered poor. Eli S. Haynes was able to calculate an orbit that matched the observations, but only with difficulty. The orbit had a period of 4.11 years and was elliptical. The asteroid's closest approach to the Sun was a distance of 1.186 AU (astronomical unit, the mean distance from the Sun to Earth). The closest approach to Earth had been in mid-September 1911, at a distance of only 0.204 AU.

More apparent observations were later found at the Heidelberg, Johannesburg, and Lick observatories. Haynes quickly realized that these could not all be of the same object. He continued to hope that at least some of them could be 1911 MT, but he was never able to prove that any were consistent with the confirmed sightings. Although 1911 MT was now lost, it received the permanent designation 719 Albert.

In the early 1970s Paul Herget and his associates at the Minor Planet Center undertook an unsuccessful effort to recover 719 Albert. Herget concluded, "The paucity of good observations and the short arc make the orbital determination a guessing game. . . . Our conclusion must be that the uncertainty of the elements is so great that the planet cannot possibly be recovered by the usual methods at this late date." The predicted positions of Albert in the two calculated orbits were 140 degrees apart—nearly opposite sides of the sky.[38]

ASTEROID MOONS?

In the late 1970s a new mystery was appearing—the possibility of asteroid moons. In 1802 William Herschel observed 1 Ceres and 2 Pallas but found no hint of a moon. Given their small sizes, Herschel concluded, "there can be no great reason to expect that they should have any satellites. The little quantity of matter they contain would hardly be adequate to the retention of a secondary body." Nearly a century later, in 1901, Charles Andre suggested that asteroid light curves were caused by the mutual eclipses of the main body and a moon.

Asteroid moons were largely forgotten until the late 1970s. Speculation on their existence returned in a roundabout way, in connection with attempts to measure the size and shape of asteroids. Even with the methods developed in the 1970s, the measurements still had a significant margin for error. The one sure way, however, was to take photometric observations as an asteroid passed in front of a star. The time that the starlight was occluded would give an exact measure of the asteroid's size. Measurements from a widely spaced group of observers would also trace out the shape of the asteroid. The shadow cast by the asteroid would be projected onto Earth's surface, and each observing site would measure a line through the shadow.

The first indication of possible asteroid moons came during the occultation of 6 Hebe on March 5, 1977. The observer was P. D. Maley. The shadow path was some 900 kilometers south of Maley's location in Texas, but he noticed a 0.5-second "secondary event." This meant that the star seemed to dim briefly, then return to its normal brightness. Maley interpreted this as the passage of a 20-kilometer moon in front of the star. The possible moon was informally named at a 1981 astronomic meeting. David W. Dunham reportedly said, "If Hebe turns out to be a binary, then we already know its name—Jebe."[39]

The most significant observation of a possible asteroid satellite came during the occultation of 532 Herculina on June 7, 1978. The observations were made by Edward Bowell and Michael A'Hearn at Lowell Observatory, James McMahon at Boron, California, and Keith Horne at Rosamond, California. The latter two had extensive experience observing occultations of stars by the Moon. The timings indicated that Herculina was 243 kilometers in diameter, about 10 percent larger than previous values. If Herculina was circular, however, then observers near Fresno, California, should have seen a brief occultation. None was seen, and it appears that Herculina's northwest edge is flattened enough so that it did not block the star.

About 97 seconds before the occultation occurred, the Lowell Observatory photometric recording showed a secondary event, which lasted 5.1 seconds. The same event was also observed visually by McMahon at Boron. The two independent observations supported the idea that this secondary event was real. Horne, farther to the west, did not see the secondary event. If it was caused by a moon, then the object was about 45 kilometers across and was 977 kilometers from Herculina. McMahon also reported seeing five other secondary events within a few minutes, lasting 0.5 to 2.1 seconds. These events were not observed by either Lowell Observatory or Horne.[40]

Although the secondary event was independently confirmed, the circumstances raised doubts. The dawn sky was brightening as the occultation occurred, and the asteroid was also very low in the sky, a mere 2.5 degrees as seen from Lowell Observatory. For those reasons, it was not possible to rule out instrument errors, atmospheric effects, or a passing aircraft.

Despite the somewhat ambiguous nature of the event, it encouraged further efforts. The first organized search for asteroid moons was undertaken on December 11, 1978, during the occultation of 19 Melpomene. One visual and three photometric observations were made of secondary events that may have been caused by several satellites.

That search was followed by the October 10, 1980, occultation of 216 Kleopatra. Teams were spread out across Oregon, Washington, California, British Columbia, and Alberta. Their results indicated that Kleopatra was roughly elliptical in shape and about 95 × 130 kilometers in size. There was also a report of a possible satellite. Gerald Rattley and Bill Cooke observed visually from sites 2,000 feet apart near Loma Prieta, California. The site was outside the path of the primary occultation, but both saw a secondary event lasting 0.9 seconds and coinciding within 0.7 seconds. They also reported a change from the blush color of the asteroid to the reddish appearance of the star during the event. No such event was recorded at Lick Observatory, which was 3.4 kilometers east of their position. One estimate, based on the secondary event, was that Kleopatra had a moon perhaps 8 kilometers in diameter and about 475 kilometers east of the asteroid.[41]

The stories of two more asteroids show the ambiguous nature of the evidence and the inability to find proof. The first was 2 Pallas. During the 1970s there were three cases of secondary occultations involving Pallas. During the May 29, 1978, occultation, for example, the 28.16-second main occultation was followed 3 seconds later by a secondary occultation lasting about 50 milliseconds. This implied a 1-kilometer satellite. Then, in 1980, speckle interferometric observations of Pallas seemed to indicate a large satellite, about 175 kilometers in diameter, which orbited at a distance of about 750 kilometers from the asteroid. The change in the positional angle implied the moon's orbital period was identical with Pallas's 7.88-hour rotational period. In 1982, however, negative evidence was published about a moon of Pallas. Radar observations of Pallas using the Arecibo dish in Puerto Rico showed no evidence of a large moon. However, even a moon as large as that suggested by the speckle observations would be near the limit of detectability of the Arecibo dish.[42]

The issue was effectively decided during the May 29, 1983, occultation of Pallas. More than 100 observers were watching Pallas from outside the occultation path for any evidence of a moon. No secondary occultations were seen, although bad weather prevented observations near the path's southern limit. Although that still left a small possibility for a moon to escape detection, it was eliminated by a reexamination of the speckle observations of Herculina. The speckle observations were best explained by a highly flattened prolate spheroid shape for Herculina, rather than by a satellite. The implication was that the earlier Pallas report was also caused by a similar misinterpretation of the asteroid's shape.[43]

The second case, of 9 Metis, was both of longer duration and more confusing. The events began with the Metis occultation on December 11, 1979, which crossed Venezuela and Guyana. A group from the Sociedad Astronomica de Venezuela observing from Barquisimeto reported seeing a secondary occultation. Dunham calculated that if the secondary occultation was actually due to a satellite, then the object might be only 1.5 magnitudes fainter than Metis itself and would be separated from the main body by a sufficient distance possibly to be visible during January through March of 1980.[44]

The report was immediately challenged. Two groups observed the occultation from Caracas and another at Apartaderos, Venezuela. None of them saw a secondary occultation. The group at Apartaderos took photos at the time of the reported secondary occultation that showed no interruption of the star's trail. They also made visual observations that showed no traces of a moon. Following the occultation, D. J. MacConnell, who was with the Apartaderos group, observed Metis on five nights between February 12 and March 7, 1980. He never suspected the presence of a satellite.[45]

Wang Sichao and his coworkers at Purple Mountain Observatory near Nanjing, China, undertook a photographic search for asteroid satellites between December 23, 1979, and February 23, 1980. Metis was photographed on thirteen different nights. Six of the photos showed a bump on the asteroid. The elongation was on the north-northeast side on three nights, on the south-southwest side on three other nights, and the asteroid appeared round on the remaining seven nights. Guiding errors were ruled out, no plate flaws were apparent, and no stars were found at Metis's position.

The observed elongations were consistent with a 4.61-day orbital period for a moon at a distance of 1,100 kilometers from the asteroid. If Metis was 150 kilometers in diameter, then the moon would be about 60 kilometers in diameter. This was consistent with estimates made from the December 11, 1979, secondary occultation, which indicated a 65-kilometer diameter for the moon and a period of 4.59 days.[46] However, Metis had a rotation period of 5.064 hours, and the difference between it and the moon's calculated orbital period of 4.61 *days* would cause severe tidal effects. For this reason, the moon's existence was doubted by most astronomers.[47]

Between mid-August and the end of November 1982, Metis was ideally placed for observations from Earth, and a search effort was organized for the satellite. The double star ADS 968 was selected as the comparison object, as the apparent separation and brightness of the two stars were identical to that calculated for Metis and its possible moon. Metis was observed by the 40-inch and 36-inch telescopes at Lick Observatory, the 24-inch telescope at Lowell Observatory, and the 20-inch telescope at Van Vleck

Observatory. Frederick Pilcher, who had proposed the search, observed Metis on many nights with his 14-inch telescope. In every case, ADS 968 was resolved as a double star, but no trace of a companion object to Metis was ever found.[48] Observations were also conducted with the 61-inch Catalina telescope, at 1,100 power under excellent seeing conditions, and later with the Multiple Mirror telescope. They also failed to show any evidence for a moon of Metis.[49]

Starting in 1984, Tom Gehrels and two colleagues began a systematic search program for asteroid moons using telescopes at Kitt Peak and the Santa Catalina Mountains. Ten of the largest asteroids were observed on the assumption that they were capable of holding onto a satellite large enough to be seen from Earth. CCD images were taken both near and far from the asteroids over several nights. Any moons larger than 2 kilometers in diameter should have been visible. None were found.

This result supported calculations that indicated an asteroid moon could not exist for a prolonged period of time. Over the lifetime of the solar system, collisions would have knocked any moons smaller than 30 kilometers in diameter out of orbit. An asteroid satellite could subsequently be formed by an impact, but the probability of such a fluke asteroid satellite was viewed as low. The fragments blasted off the asteroid would have to be going at a very narrow speed range to go into orbit around it. Too low a velocity would cause them to fall back, and too high a speed would cause the fragments to escape from the primary body.[50]

Although a few secondary occultations of asteroids were reported in the 1980s and early 1990s, the number dropped sharply after the failed Metis search. For more than a decade, sightings had been reported, but none provided unambiguous proof of asteroid moons. The repeated failure to find proof only served to fuel doubts that they existed. Yet, double craters on Earth, the Moon, and other planets continued to hint at the existence of binary asteroids.

HUBBLE OBSERVATIONS OF VESTA

The culmination of the modern era of asteroid research was the observation of asteroids by the Hubble space telescope. Although primarily a deep-sky telescope, it also had the capability to make planetary observations. Between 1990 and 1993, Hubble observed a dozen asteroids. Because of its flawed mirror, Hubble could barely resolve them, and little could be learned. Following a shuttle repair mission, however, the Hubble began to perform as designed.

Between November 28 and December 1, 1994, Ben Zellner and his colleagues used the Hubble to take images of 4 Vesta in four different wavelengths. One series covered nearly one full revolution and showed its flattened spherical shape and surface markings. The Hubble images surpassed those taken by the European Southern Observatory's 3.6-meter telescope. Even though the 3.6-meter telescope had a larger mirror and adaptive optics that compensated for atmospheric effects, the Hubble was still able to resolve finer details.[51]

The Hubble observations were used by Zellner to produce maps of both Vesta's surface and its composition. The surface maps showed a dark, circular feature, which might be a crater some 200 kilometers across. It was dubbed Olbers, after the asteroid's discoverer. The composition map was more striking. The map indicated that one hemisphere was covered with lava flows, which were believed to be the asteroid's ancient crust. The other hemisphere, in contrast, had the spectral signature of once molten rock that had cooled and solidified deep underground. The rock was then exposed by subsequent impacts. The Hubble data expanded on previous ground-based observations of Vesta's composition, and they confirmed that Vesta was also the source of eucrite and diogenite meteorites.[52]

In May 1996 Vesta was closer to Earth than at any time in the previous decade. Its angular size was 50 percent larger than it had been in November-December 1994, and Vesta's south pole was better placed for observations. Both Zellner and Peter C. Thomas of Cornell University used Hubble to image Vesta. The results were better than the previous observations. Zellner was able to produce an elevation map showing surface details. Thomas produced 3-D computer images that made Vesta appear to be a potato-shaped object in space.

The major discovery was a huge crater close to Vesta's south pole. It was some 640 kilometers in diameter and an average of 6 kilometers deep. A 13-kilometer-high central peak loomed above its floor. It had been produced when the molten rock sloshed back into the crater after the impact. The crater was about 85 percent of the diameter of Vesta and had a volume of roughly a million cubic kilometers. The volume of material lost was so great that it was believed that Vesta's rotation axis had been shifted, with the crater ending up at the south pole. The debris blasted out had formed the two dozen 5- to 10-kilometer asteroids that made up the Vesta family of objects. Some of the debris had subsequently gone into orbits that traveled out to a point some 2.5 AU from the Sun. Here it was affected by the gravitational forces of Jupiter and sent into highly elliptical orbits. Some of the material subsequently became the eucrite, diogenite, and howardite meteorites recovered on Earth.

Although the Vesta family and meteorites implied the asteroid had suffered a massive impact sometime in the distant past, the feature itself was something of a surprise. Thomas said later, "In hindsight we should have expected finding such a large crater. But it's still a surprise when it's staring you in the face."[53] The crater also confirmed the previous ideas about the evolution of asteroids early in the history of the solar system, and the formation of asteroid families. If the object that struck Vesta had been larger, then the asteroid would have been shattered and a more extensive family would have been formed from the debris. It was a fate that had befallen the other large asteroids.

4

APOLLOS, AMORS, ATENS, AND CLOSE CALLS: THE NEAR-EARTH ASTEROIDS

Asteroid 1994 MX1 made its closest approach at 2:15 p.m., approximately 14 hours after [James V.] Scotti's first observation, when it was above the Northern Hemisphere and 65,000 miles away. Scotti said if it had struck the Earth, the asteroid probably would have landed in Russia.

Washington Post, December 12, 1994

THE FIRST INDICATION THAT asteroids might travel outside the main belt came in the early morning hours of June 14, 1873, when James C. Watson discovered 132 Aethra. The asteroid could be followed for only three weeks before it was lost. That was sufficient, however, to calculate a preliminary orbit, which showed 132 Aethra crossing the orbit of Mars. More than 49 years would pass before Aethra was accidentally recovered. The next discovery outside the main belt was 433 Eros on August 13, 1898, by Gustav Witt. When its orbit was calculated, the results were a surprise: Eros came within 1.13 AU of the Sun, under ideal conditions it would come as close as 14 million miles to Earth, and the farthest point of its orbit was well outside that of Mars.

It was soon apparent that Eros was not alone. In 1911 the asteroid 719 Albert was discovered. Although it was soon lost, the rough orbit indicated not only that it approached Earth but also that it possibly ranged out as far as Jupiter's orbit. Two other asteroids, 887 Alinda and 1036 Ganymed, discovered in 1918 and 1924, had orbits similar to the long-lost Albert. In 1931 Eros made a well-observed close approach to Earth. The following year would see the first of a series of asteroids that came even closer.

AMOR, APOLLO, ADONIS, AND HERMES

The first of these near-Earth asteroids (NEAs) was discovered on March 12, 1932, by the Belgian astronomer Eugene Delporte. Named 1221 Amor, its closest approach to the Sun was only 1.08 AU, closer than Eros. Amor also passed only 10 million miles from Earth. A small object, only about a half-mile in diameter, it was then the faintest asteroid to be given a permanent number. Despite its small size, Amor has been well observed on its subsequent close approaches, every eight years.

Amor held the distinction of being the asteroid that came closest to Earth for a mere six weeks. On April 24, 1932, Karl Reinmuth of the Heidelberg Observatory in Germany spotted the long trail of an asteroid on a photographic plate. Given the initial designation 1932 HA, it was followed until May 15, when G. Van Biesbroeck photographed it from Yerkes Observatory. That night, 1932 HA passed only 7 million miles from Earth. Its motion soon caused it to pass into the evening twilight, and it was lost for 41 years.

The preliminary orbit that was calculated crossed the orbits of both Earth and Venus. Its closest approach to the Sun, called the perihelion, was on July 7, at a distance of only 0.65 AU. The orbit also indicated that if the perihelion date had been four days earlier, the asteroid would have passed only 3 million miles from Earth. This asteroid was estimated to be about two or three times larger than Amor. Because its orbit carried the asteroid so close to the Sun, Reinmuth gave it the name Apollo, after the god of the Sun.[1]

These two asteroids became the prototypes of separate classes of asteroids. An Amor-type asteroid was defined as having a perihelion distance of between 1.017 AU and 1.3 AU. The asteroid makes close approaches to Earth but never crosses Earth's orbit. An Apollo-type asteroid, in contrast, has a perihelion distance of less than 1.017 AU. Because the most distant point of Earth's orbit is 1.017 AU from the Sun, the asteroid is considered to cross Earth's orbit. Because of gravitational effects on the orbits, the perihelions of some Amor asteroids oscillate across the 1.017 AU boundary. Asteroids that do not come as close as 1.3 AU are simply referred to as Mars-crossers.[2]

Apollo held the record for a close approach to Earth for just under two years. On February 12, 1936, Delporte spotted 1936 CA. Only a few days before, it had passed within 1.5 million miles of Earth. Its orbit was better placed for observation than Apollo, and it remained visible in the night sky; however, Adonis (as the asteroid was subsequently named) was much dimmer than Apollo. Within two months of its discovery, Adonis could be observed only with the 100-inch telescope on Mount Wilson, then the largest in the world. Adonis was lost for the next 41 years.

This series of remarkable asteroids was ended by Reinmuth when he discovered 1937 UB. On October 30, 1937, it passed within 500,000 miles of Earth, only twice the distance between Earth and the Moon. The asteroid passed into the daytime sky before word could be spread of its discovery. Images of the asteroid, later named Hermes, were found on photographs taken at Harvard University, Sonneberg, Germany, and Johannesburg, South Africa. They covered a span of less than five days, however, and Hermes was considered hopelessly lost.

SUBSEQUENT DISCOVERIES

For a decade after this initial series of discoveries, Amor, Apollo, Adonis, and Hermes remained unique curiosities. No other NEAs were discovered, and asteroid research itself was virtually halted by the war and the postwar disruption. Like the series of the 1930s, subsequent discoveries of Apollo asteroids would be clustered in time, with long intervals between them. The drought was broken on December 12, 1947, with the discovery of 1947 XC, by Henry Gicles at Lowell Observatory. The asteroid was identified on two plates taken by Gicles, but the discovery could not be followed up. It remained lost for another 32 years.

The most successful discoverer of the 1947–1951 series was Carl A. Wirtanen. He was not searching for NEAs, however. Rather, he was undertaking a proper-motion study, using the 20-inch astrograph at Lick Observatory, to measure the change in position of nearby stars as they traveled through space. In the process, he found three new Apollo-type asteroids. The first was 1948 EA, spotted on March 7, 1948. Like most of the other NEAs, it was subsequently lost.

It was soon followed by 1948 OA, found on July 17. Unlike its luckless predecessor, 1948 OA was observed well enough to allow recovery on its close approaches to Earth in 1956 and 1964. It was later named 1685 Toro. (It was then standard practice among astronomers to give asteroids with unusual orbits four-letter, masculine names, rather than the female names traditionally given to asteroids.)[3]

Toro proved to have a remarkable orbit, ranging from a maximum distance from the Sun of 1.96 AU to as close as 0.77 AU. Toro crossed the orbits of both Mars and Earth and came within 0.05 AU of the orbit of Venus. Toro's orbital period of 584.2 days also meant that for every five revolutions Toro completed, Earth made eight and Venus thirteen revolutions. Toro is thus in resonance with both planets. Toro makes close approaches to both planets at regular intervals, but this resonance also prevents it from colliding.[4] The third of Wirtanen's asteroids was 1950 DA. Like his first, its orbit was uncertain, and the asteroid was lost.

The other two Apollo asteroids discovered during this period were found by astronomers using the 48-inch Schmidt telescope at Palomar Observatory. The first was the most remarkable and important of the early NEAs. On the night of June 26, 1949, Walter Baade was using the new telescope to take a one-hour exposure of the sky near the star Antares. When the plate was developed, it showed the long trail of an NEA.

Originally called 1949 MA or, less formally, Baade's object, the asteroid has several unique properties. At perihelion it passes twice as close to the Sun as Mercury, but its aphelion (the farthest point from the Sun) is 2 AU, beyond the orbit of Mars. More significant, 1949 MA's orbit is inclined 23 degrees to the ecliptic. The orbit is more like that of a comet than an asteroid. At the time of its discovery, 1949 MA was passing about 4 million miles from Earth. The asteroid also makes seventeen revolutions for every nineteen made by Earth, meaning it would make another close pass in 1968. The asteroid's orbit was observed well enough to allow recovery, and it was named 1566 Icarus. Like its mythological namesake, Icarus also flew too close to the Sun; at its closest approach, the asteroid's surface would actually glow from the heat.[5]

On September 14, 1951, the second discovery of an Apollo-type asteroid was made with the 48-inch Schmidt. A. E. Wilson and R. Minkowski discovered 1951 RA, subsequently named 1620 Geographos. Like

Icarus, Geographos approached close to Earth, passing some 6 million miles away in 1969. The discovery of Geographos brought the second series of Apollo discoveries to a close.

It would be the end of the 1950s before the third series of Apollo asteroids was discovered. The first, 1959 LM, was found by Cuno Hoffmeister on several plates taken at Boyden Observatory in South Africa. He did not find the images until the next year, however, so the asteroid was lost. During September and October 1960, the Palomar-Leiden asteroid survey used the 48-inch Schmidt to take a census of asteroids. Two Apollo asteroids were spotted on the plates, 6344 P-L and 6743 P-L. Although a rough orbit could be calculated from the images, it was not precise enough to allow recovery. A third Apollo asteroid was found on the Palomar-Leiden plates, 5025 P-L, but it was not identified until many years later and, like the other two, was lost.

The fourth series of Apollo asteroid discoveries was made between 1971 and 1973. The first two were 1971 FA and 1971 UA, discovered on March 24 and October 26. They were followed by 1972 XA, discovered December 5, 1972, by Paul Wild, and noted both for its size (estimated at 4 miles in diameter, the largest Apollo asteroid) and for its orbital inclination of 41 degrees (almost double that of Icarus). This high inclination was the reason that 1972 XA escaped detection for so long. It spent only a small part of its orbit near the ecliptic and celestial equator, where most asteroid searches are made. Prediscovery images of 1972 XA were later found on films taken by Diethard Ruhnow and 15-year-old Stefan Petermann at a school observatory in Rodewisch, East Germany. The photos had been made on November 25, when the asteroid was 10 million miles from Earth and had a brightness of magnitude 9.[6] All three asteroids were recovered; they were named 1864 Daedalus, 1865 Cerberus, and 1866 Sisyphus.[7]

The year 1973 saw two more Apollo discoveries. The first, 1973 EA, was found on March 6 by Charles T. Kowal using the 48-inch Schmidt telescope on Palomar Mountain. Observations by Eleanor Helin four nights later with the 18-inch Schmidt on Palomar showed the new asteroid had a 40-degree inclination. At this point, the full moon intervened, making observations impossible. Once the full moon had passed, Paul Wild resumed observations. Although it was entering the evening twilight by mid-April, enough observations had been made to allow a successful recovery. It was then named 1981 Midas.

Helin then discovered 1973 NA on the Fourth of July with the 18-inch Schmidt. The object was moving rapidly southward, but she was able to photograph it on July 6, 7, and 8. After that, 1973 NA was too far south to be observed from Palomar. Carter Observatory in Wellington, New Zealand, made additional observations. When a rough orbit was calculated, it showed that 1973 NA had an inclination of 68 degrees, higher than any other Apollo asteroid.

During this same period, about a dozen Amor-type asteroids had also been discovered. Among them was 1953 EA, discovered on March 9, 1953, by A. E. Wilson with the 48-inch Schmidt. More than three months later, it was observed by Allan Sandage with the 200-inch Hale telescope. The orbit calculated for 1953 EA proved unusual, with a period of nearly four years, meaning it made close approaches to Earth at this interval. The four-year period also put it in resonance with Jupiter, as 1953 EA made three revolutions for every one made by Jupiter. As a result, 1953 EA came no closer to Jupiter than 1.7 AU, which prevented catastrophic gravitational effects. However, it also pointed out the instability of Amor and Apollo orbits. Before 1940 the asteroid had been in an Apollo-type orbit. In the 1870s its perihelion distance had been as low as 0.94 AU. Despite the number of observations of 1953 EA, the object was lost for the next 20 years.[8]

THE ATEN-TYPE ASTEROIDS

Before the early 1970s the discoveries of Apollo and Amor asteroids had been accidents. Now, deliberate searches for NEAs were beginning. Asteroid 1973 NA was the first discovery of the Palomar Planet-Crossing Asteroid Survey by Eugene Shoemaker and Eleanor Helin. New Apollo and Amor asteroids began to be discovered on a regular basis. During the 1970s two or three new Earth-crossing asteroids were spotted each year. The search effort also led to the discovery of a third type of NEA.

During the early years of the program, Helin and Shoemaker observed with the 18-inch telescope for four nights every lunar month. The film from the night of January 7, 1976, showed an asteroid trail. The asteroid, designated 1976 AA, was the first to be discovered in 1976. Estimates were made of its position for the next night's observations, when ten more exposures were made of the asteroid. During the day of January 9,

Helin's assistant, Schelte J. "Bobby" Bus, made position measurements, and a preliminary orbit was calculated by James G. Williams of the Jet Propulsion Laboratory.[9]

Over the nights of January 9–12, additional observations of 1976 AA were made not only at Palomar but also at Harvard Observatory, Table Mountain Observatory, Mount Wilson Observatory with the 60-inch telescope, and the Jakiimo and Nihondaira observatories in Japan. One amateur astronomer, Steven Wheatcraft of Coshocton, Ohio, using an 8-inch telescope, also photographed 1976 AA. On January 10 and 11 Bus made measurements of the asteroid's position, which were then used by Williams and Brian Marsden to calculate an orbit.

Their orbit was unlike that of any Apollo or Amor asteroid previously discovered. At perihelion, its distance from the Sun is only 0.791 AU, slightly outside the orbit of Venus. The aphelion distance is 1.141 AU, just outside the orbit of Earth. The period is 346.8 days, making this the first asteroid to have a period of less than one year. Because of its unique orbit, 1976 AA became the first of a new type. It was given the designation 2062 Aten, after the Egyptian god of the Sun. Aten-type asteroids are defined as having periods of less than one year and aphelion distances of 0.983 AU or greater (0.983 being the perihelion distance of Earth's orbit).

Initial spectrophotometric studies at the University of Arizona and University of Hawaii indicated that Aten was a C-type asteroid, a very dark object similar to carbonaceous chondrite meteors. Its diameter was originally estimated to be about two miles.[10] This announcement was quickly retracted when additional observations indicated Aten had an albedo of about 21 percent. It was later determined to be an S-type asteroid, about half a mile in diameter.[11]

Aten-type asteroids are the rarest of the NEAs. By March 1986, after a decade of searching, only six were known: 2062 Aten, 2100 Ra-Shalom, 2340 Hathor, 3362 Khufu, 3554 Amun, and possibly 1954 XA (a lost object). All of them had been discovered at Palomar, and five of the six with the 18-inch Schmidt.[12] A seventh Aten, 3753 1986 TO, broke this monopoly. It was discovered on October 10, 1986, at Siding Spring Observatory in Australia. As of mid-1987 the breakdown of NEAs was 46 Apollos (7 lost), 47 Amors (3 lost), and 7 Atens (1 lost). Another 73 asteroids were in Mars-crossing orbits.[13]

It was another three years before the next two Atens were found. (In contrast, the number of Apollos had now grown to 63.) The first was 1989 UQ, found on October 26 at Cote d'Azur Observatory, France, by C. Pollas. Then, only a week later, on November 2, Eugene Shoemaker, his wife, Carolyn, and their assistant David H. Levy found 1989 VA. It had the distinction of being the Aten with the shortest period, only 227 days, nearly the same as the planet Venus. (The longest-period Aten was 3753 1986 TO, at 364 days.)[14]

THE HUNT FOR THE LOST APOLLOS

NEAs pose observing problems much greater than those of main-belt objects. They are small, meaning they can be seen only when they are close to Earth. Because they are close to Earth, they move through the night sky at a high angular rate and can be observed for only a limited time, making it difficult to calculate a reliable orbit. Because of the elliptical orbits of these objects, it may be up to two decades before they are again well placed for recovery. For these reasons, most of the early Apollo-type asteroids were lost. With the resurgence of asteroid research in the early 1970s, efforts began to recover the lost Apollos.

Since Apollo had been last seen in April and May 1932, it had made close approaches to Earth in 1939 and 1941. Not until 1971 and 1973 was a systematic hunt undertaken. The most probable orbit for Apollo indicated that the asteroid would be near Earth, but its exact position was not known. All that could be determined was that it would be somewhere along a line in the sky on each specific night. The first try was made in 1971 by Charles Kowal at Palomar and Paul Wild at Berne University. Although they took a series of photographs along the expected track, Apollo was not found. Apollo was apparently fainter than had been believed, and it was determined that a wider area should be covered in the 1973 attempt.

In February 1973 Richard E. McCrosky decided to attempt to recover Apollo. For more than a year he had used Harvard University's 61-inch telescope to observe faint comets and unusual asteroids. He decided that the best time to search was in March and early April, as Apollo's motion and the uncertainty of its position would be minimal. The calculated positions were based on assumed dates of perihelion, ranging from 30 days before the estimated date of July 10 (T–30), to 30 days after (T+30). The result was a search area in the form of a narrow strip of sky more than 90 degrees long. To make matters more difficult, the 61-inch telescope was ill suited for a wide-area search. Each photographic plate could cover an area of only 40 × 50 minutes of arc, requiring more than 100 plates to cover the search area.

McCrosky's search efforts were further delayed by bad weather and then the full moon. It was not until March 27 that McCrosky and Cheng-Yuan Shao could begin. The T+30 part of the search area was the first to rise above the horizon, and McCrosky began the first 25-minute exposure. The procedure he used was the same as that developed by Joel H. Metcalf at the turn of the century. McCrosky guided on the calculated position of Apollo, rather than the stars, so that the stars would trail while Apollo would appear as a spot. When the exposure was completed, McCrosky went to the darkroom to develop the plate.

McCrosky finished developing the plate and put it on a light frame. Amid the star trails, within three-quarters of an inch of the center, was a single spot. A second exposure was made of the same area, and it showed the object had moved the calculated distance. Apollo had been found on the very first exposure of the search program, after being lost for 41 years. A third plate was taken that night and more on the following two nights. All showed Apollo, and there was no longer any doubt. In early April, Apollo was also observed by Elizabeth Roemer with the 61-inch and 90-inch telescopes of the University of Arizona. Having been recovered, it was finally given a permanent number, 1862 Apollo.[15]

Although the orbit of Adonis was considered to be more reliable, recovery would still be difficult. Adonis was a much smaller object than Apollo and so could be seen only during a close approach. The first attempt at recovery, in 1943, was unsuccessful. Adonis passed some 20 million miles from Earth in 1959, but it was not seen. The difficulty of recovery was compounded because Adonis passed within 0.04 AU of Venus in 1964.

The next chance came in 1977. Adonis could be anywhere along a line stretching some 50 degrees. On the night of February 14, 1977, Charles Kowal took ten plates with the 48-inch Schmidt to cover the search area. One of the 8-minute exposures showed an asteroid trail with a length close to that calculated for Adonis. Word of the discovery was sent to H. E. Schuster at the European Southern Observatory. He then was able to photograph Adonis with the 1-meter Schmidt on five straight nights, beginning on February 24. Like Apollo, Adonis had been missing 41years. It was given the permanent designation 2101 Adonis.[16]

Over the next two decades, several more of the missing Apollos were recovered. The Apollo asteroid 1947 XC, the first postwar discovery, was recovered by Helin on December 13, 1979, at Palomar, 32 years and one day after it was first observed. Later it was named 2201 Oljato. Another of the Apollos discovered and then lost during the 1947–1951 series was 1948 EA. It was also recovered and was designated 1863 Antinous. Of the 1959–1960 series, two of the four were recovered. The asteroid 1959 LM was designated 4183 Cuno, after the first name of its discoverer, Cuno Hoffmeister. One of the three Apollos found during the Palomar-Leiden Survey, 6743 P-L, was also recovered. Marsden later said that he "sweated blood" over the identification. The asteroid was given the number 5011. Still among the missing are Hermes (the third Apollo asteroid discovered), 1950 DA, 1954 XA (a suspected Aten), and 5025 P-L and 6344 P-L.[17]

APOLLO ORIGINS

An understanding of the origins of the NEAs has proven elusive. The first to undertake a study of the Apollo asteroids was Ernst Opik of Armagh Observatory in 1951 and 1963. He began by assuming the handful of Apollo asteroids known had occupied their present orbits since the formation of the solar system 4.5 billion years ago. He further assumed that the rate of collision was higher in the past, that the rate of collision was proportional to the total number of Apollo asteroids, and that the collision rate with Earth was once per 100 million years. Based on that, Opik calculated that for the few surviving Apollos to still exist, there would have had to have been several trillion Apollo asteroids at the time of the solar system's formation. Their total mass was 1,000 times the mass of the Sun.[18]

This result was so absurd that it was clear to Opik that the NEAs must be short-term visitors to the inner solar system. There would have to be a constant resupply of new Apollo asteroids to replace those lost in collisions. Opik could not think of a mechanism that would send main-belt objects into the inner solar system, however. He proposed that the Apollo asteroids were actually comet nuclei that had lost their gas and dust.[19]

The same new observation techniques used on main-belt asteroids have also been used on near-Earth objects. The results have been mixed as to their origin. Of a group of 24 NEAs, 2 are carbon-rich C-type asteroids, 12 are S-type stony objects, 1 is a U-type, and 1 is "other." Since comets were formed in the outer reaches of the solar system, in the icy depths beyond Pluto, they would not be expected to be stony bodies.

Lucy McFadden made 24-color spectrophotometry observations of seventeen NEAs in the early 1980s. She concluded that although they were composed of the same materials as main-belt asteroids, the NEAs had more silicate materials. Strong ultraviolet and near-infrared absorption bands were present in 75 percent of the near-Earth objects, but only 40 percent of main-belt asteroids showed those features. McFadden also proposed that the absorption features were due to surface texture, although the specifics were not known. Both these factors pointed to the NEAs' being relatively new objects, fragments of older asteroids that were blasted off in collisions, with fresh, unweathered surfaces. This was consistent with the data on asteroid shapes, which indicated that NEAs are more irregular compared to main-belt objects.

While McFadden's observations suggested that NEAs were fragments of main-belt objects, data on their rotation rate pointed to a cometary origin. Of 22 Apollo-Amor objects, most had rotation rates between 2.27 and 6.8 hours. A few had slightly longer rates, between 8 and 10.19 hours, and a few had substantially longer rates, between 19.79 hours (2100 Ra-Shalom) and 148 hours (3102 1981 QA). The split between asteroids with very short rotations and those with much longer rates implied two distinctive types of NEAs. Although the number of objects was too limited to draw statistical conclusions, this split was similar to the rotation rates found in comets, bolstering the idea that at least some were extinct comets.[20]

An unexpected discovery about Apollo asteroids came from the IRAS (infrared astronomy satellite) mission. IRAS was launched on January 25, 1983, from Vandenberg Air Force Base. It carried a 22.4-inch telescope cooled to −455 degrees F by liquid helium. IRAS was to conduct a survey of nearly the entire sky in infrared wavelengths. Although the mission's primary goal was stellar astronomy, the IRAS data were also being used by John K. Davies and Simon Green of the University of Leicester to search for moving objects such as asteroids and comets.[21]

On October 11, they noticed a fast-moving object in seven consecutive scans. The object was an Apollo asteroid, designated 1983 TB. The following night Charles Kowal photographed it with the 48-inch Schmidt. The orbit of 1983 TB showed how remarkable an object it was. Its perihelion distance was only 13 million miles, one-third the distance of Mercury and well inside the orbit of Icarus, the previous record holder. At the closest approach, the asteroid glowed red. The orbit's farthest point was outside the orbit of Mars, and the asteroid had a relatively short period of 1.43 years. In recognition of its close approach, it was given the name 3200 Phaethon, after the son of Apollo, who drove his father's chariot so close to Earth that it scorched the equatorial regions.

Almost immediately, Fred Whipple realized Phaethon's orbit was close to that of the Geminid meteor shower, which appears every December 6–17 and is one of the year's best showers. It had been known since the mid-nineteenth century that meteor showers were caused by the debris of comets, but no comet had ever been identified as the source of the Geminid shower. The discovery that the Geminids were from an asteroid was unexpected, but it fit with Opik's idea that Apollo asteroids were extinct comet nuclei. A problem was that although Phaethon's orbit was unlike that of most asteroids, it was also unlike that of any known comet.[22]

Phaethon made a close approach to Earth in December 1984, and the results were confusing. Initial observations seemed to indicate it was an S-type asteroid. Subsequent observations by David Tholen from Mauna Kea Observatory indicated that Phaethon was somewhat bluish, which was inconsistent with an S-type spectrum. A group of British astronomers observed Phaethon at infrared wavelengths and determined that it did not resemble an S-type asteroid or any other type. It was also different in appearance from the comet Neujmin 1, which was then inactive and asteroidlike in appearance.

The British astronomers also measured the asteroid's heat emissions, which indicated that Phaethon's surface conducted heat like bare rock and that it was rotating rapidly. This was consistent with Tholen's results, which indicated a rotational period of four hours. Both results were inconsistent with an extinct comet. Data from the Halley's comet flyby missions later showed it was a very dark object with a thick crust of carbon-rich materials. Such material is a good insulator, in contrast to rock. The high rotation rate was also inconsistent with a comet nucleus, thought to be a fragile collection of ice and dust that would tend to break apart if spun so rapidly.[23]

The Geminid meteors also seemed to point to a stony parent body. The meteors in a typical shower have a density of about 300 kilograms per cubic meter, consistent with light, fluffy, dustlike material from a comet. In contrast, the Geminid meteors have a density of nearly 1,000 kilograms per cubic meter, as would be expected from a stony body. Overall, the data are consistent with the idea that Phaethon is an S-type asteroid, rather than an extinct comet nucleus. The odd spectrum could be the result of its regular baking during the close approaches to the Sun.[24]

After Phaethon was identified as the source of the Geminid meteors, other asteroids were also connected with meteor showers. Icarus was identified as the source of the Arietids, a major daylight shower detected by radar. The long lost Hermes was associated with the October Cetids shower. Adonis was suggested as the parent body of three showers, the Chi Capricornids (another daylight shower), Scorpioid-Sagittarids, and Capricornid-Sagittarids.[25] The asteroids 2201 Oljato, 2212 Hephaistos, 1984 KB, 1982 TA, and 5025 P-L all contribute parts to a complex system of meteor showers, the major component of which is the Southern Taurids. The asteroids associated with meteors are all Apollo-types. No associations with Amor asteroids were noted.[26] However, the meteors in two other showers, the January Bootids and the Epsilon Aquilids, have orbits indicating their parent body might be in Aten-type orbits.

In the early 1990s two astronomers independently discovered asteroid streams. Jack D. Drummond of the University of Arizona compared the orbits of 139 NEAs, while Ian Halliday and two colleagues at the Herzberg Institute of Astrophysics in Ottawa, Canada, examined the orbits of 89 fireballs photographed by camera networks across North America. Both studies looked for close similarities in orbital distances, inclinations, eccentricities, and directions. Drummond found three groups of Amor-type asteroids with similar orbits. At the same time, Halliday found four groups of fireballs with similar orbits. When the two sets of groups were compared, two of Halliday's groups closely matched Drummond's groups, and a third was a looser match. Meteor streams consist of small particles, but these asteroid streams are composed of much larger bodies. Unlike the situation with meteor showers, none of the groups were Apollo asteroids, as their Earth-crossing orbits are unstable and the grouping would disperse after only 10,000 years.

This does raise the question of how a solid body like an asteroid could produce a meteor shower or asteroid stream. The obvious conclusion is that it is the result of a collision with another body. All of the groups found by Drummond and Halliday showed minimum orbital separations at perihelion, indicating that the parent objects broke up at this point. In the case of Phaethon, however, the distribution of the meteors indicates it most likely occurred when the asteroid was at its greatest distance from the Sun.[27]

Although Phaethon is now regarded as an asteroid, there remains evidence that some NEAs originated as comets and even now display cometlike features. After its recovery, 2201 Oljato (1947 XC, the fourth Apollo discovered) soon proved to be a unique object. Oljato's orbit is highly elliptical, and the object appears to be very irregular in shape. The real surprise is its spectrum, which shows a peak at about 6,000 angstroms, along with strong ultraviolet reflections. There are also hints that its spectrum is variable. These spectrum features can be matched in the laboratory only by very unusual materials. No other asteroid or meteorite shows a similar spectrum.[28]

Lucy McFadden determined that when the features are interpreted as emission bands, rather than absorption bands, the spectrum can be interpreted as the light scattering from particles being lost from the surface of Oljato. In effect, Oljato is a comet nucleus, although its activity is too low to form a normal coma and tail. The few remaining pockets of ice and gas vent through fissures in the crust, to form a thin haze around the object. Although this fits the spectrum data, one problem is that Oljato's albedo is around 30 to 40 percent, rather than the 5 percent of a comet nucleus or the 15 to 20 percent of most NEAs.[29]

Even more remarkable is the case of the Apollo asteroid 1979 VA, subsequently given the number 4015. It had been discovered by Helin on November 15, 1979. During the summer of 1992, Ted Bowell of Lowell Observatory calculated its position backward in time to locate prediscovery images on the first Palomar Sky Survey plates. This indicated that an image of the asteroid would appear on plates taken on November 19, 1949. When Bowell examined the plates, he was surprised to discover not an asteroid trail, but a comet. Brian Marsden learned of the discovery and recalled that comet Wilson-Harrington had been cataloged from the same plates. Palomar survey plates taken after its discovery showed the comet as pointlike.

Further research showed that 4015 1979 VA and comet Wilson-Harrington were the same object and that the tail was not a plate defect. When 4015 1979 VA reached perihelion in August 1992, however, no trace of cometary activity was detected. The consensus of astronomers was that the 1949 plates had captured "the last gasps of a dying comet."[30] This comet-turned-asteroid was later named 4015 Wilson-Harrington.

The current opinion is that the bulk of NEAs had their start as main-belt objects. Their journeys began when an impact sent fragments into a resonance orbit with Jupiter. As a result of Jupiter's gravitational effects, their orbits became more eccentric. The process would take only a few hundred thousand years. Even with the low rate of collisions between main-belt asteroids, about 100 new kilometer-sized asteroids would be sent into the inner solar system every million years, a rate sufficient to replace the NEAs lost over time.

The orbits of comets can also be affected by the gravity of Venus and Earth. Over the course of many passages around the Sun, the radius of their orbits would be reduced until they no longer crossed the orbit of Jupiter, making them immune from that planet's gravitational effects and leaving them in asteroidlike orbits. The primary clue to their origin as comets would be a highly elliptical or highly inclined orbit.[31]

In his early studies, Opik had assumed that NEAs would eventually be swept up by the planets. Three decades later, computers allowed calculations of the past and future movements of asteroids. Paolo Farinella and his team at the University of Pisa calculated the orbital evolution of 47 Earth-crossing objects (both asteroids and comets) spanning 2.5 million years. The study produced a surprising answer. Farinella and his group concluded that an NEA is actually about 10 times more likely to fall into the Sun than to hit a planet.[32]

RADAR STUDIES OF NEAR-EARTH ASTEROIDS

One of the most important means of studying NEAs originated a mere four months after the end of World War II. On January 10, 1946, a U.S. Army radar successfully bounced signals off the Moon. Called Project Diana, this was the birth of radar astronomy. Optical astronomy was limited to passive observation of an asteroid's brightness, measuring any changes in that brightness as it rotates, and observing how light is absorbed and reflected from its surface. Radar astronomy, in contrast, allows us to reach out and touch an asteroid. The radar signals are transmitted at specific wavelengths and waveforms, with known power levels. How the signal's known qualities are affected by being reflected from the asteroid provides data on the object's physical nature.

The simplest information radar astronomy can provide is the asteroid's orbit. Its distance is determined by timing the round trip of the echoes, then multiplying by the speed of light. Because the asteroid is moving through space, the frequency of the echo will be shifted to shorter or longer wavelengths (called the Doppler shift). Once its distance and speed are known, an asteroid's orbit can be determined with greater precision than by optical means. Just as the asteroid is moving through space, it is also rotating on its axis. Some of the echo is reflected from areas of the asteroid turning toward Earth, other parts are reflected from areas turning away. This spreads out the frequencies of the echoes. The faster the asteroid spins, the broader the spread from the single original wavelength. The size of the asteroid also affects the spread of the echo's bandwidth.

The average surface roughness and composition of the asteroid affect the echo as well. The radar signal is described as in phase, or coherent, meaning all the peaks and valleys of the waves within the signal are the same, as if they are soldiers marching in step. If the asteroid's surface is smooth, then the coherence of the signal will tend to be preserved. If the surface is rough, then the signal will tend to scatter from the individual objects, and an incoherent echo will be produced. Both qualities are readily apparent at the receiver. Different materials will also affect the echo differently. The metallic surface of an M-type asteroid is a good radar reflector; it will produce a stronger echo, and the asteroid will be described as having a large radar cross section. A stony asteroid covered with an insulating blanket of debris will be a poor reflector, show a weak echo, and be said to have a low radar cross section.[33]

The close approach of Icarus in June 1968 was the first opportunity for radar observations of an NEA. It was one that occurred only every nineteen years, and it was beset by difficulties. Three groups tried: the Arecibo Ionospheric Observatory (which was unsuccessful), the Jet Propulsion Laboratory's Goldstone Mars Station, and the Haystack Observatory operated by MIT.

Radar observations at Haystack began on the morning of June 12,

1968, but rain interfered. For this reason, it was not until the afternoon of June 13 that an echo was detected from Icarus. Another successful run followed that evening, confirming that echoes had been detected. Additional observations were made on June 14, and the final ones were made during the morning of June 15, the day after the closest approach. Haystack was thus the first to detect an echo from an asteroid.[34]

The JPL team under Richard M. Goldstein also detected echoes from Icarus on June 14–16. The signals indicated that the surface was rough, even jagged. Icarus appeared to be shaped like a peach stone. The radar was unable to determine if the asteroid's surface composition was stony or iron. This, in turn, prevented exact size measurements. The apparent size of Icarus was just over a half-mile, a reduction of nearly 50 percent from the preradar estimate.[35]

Radar astronomy of asteroids remained limited for the next decade. Following the observations of Icarus in June 1968, it was not until 1972 that the next asteroid, 1685 Toro, was observed by radar. Using both radar echoes and optical data from the Goldstone Mars Station, Goldstein was able to determine that Toro had an irregular rocky surface smoothed by a layer of loose material.

Next was 433 Eros, in January 1975. The observations provided the best understanding of the small-scale roughness of an asteroid's surface that was obtained until the late 1980s. Goldstein and Roy Jurgens were able to determine that the surface of Eros was much rougher than the Moon or any of the terrestrial planets. Eros was completely covered with sharp edges, pits, subsurface holes, and embedded rocks.

During the 1975 close approach of Eros, Don Campbell and Gordon Pettengill also observed it at Arecibo. Campbell and Pettengill measured its radar cross section, and like the Goldstone researchers, they found its surface was rough. Although Campbell and Pettengill could not determine its specific surface material, they did conclude it could not be a highly conductive metal. The Eros observations marked both the first successful detection of an asteroid by Arecibo and its emergence as the center for asteroid radar research.

The next attempt was more ambitious: the radar detection of the main-belt asteroids 1 Ceres and 9 Metis. Pettengill, working with Brian Marsden, Goldstein, and Tom Gehrels and Benjamin Zellner of the University of Arizona, made the attempt in December 1975. They failed to detect any echoes from either of the asteroids. With Metis, the failure was no surprise—it was a small object. Ceres, in contrast, was the largest asteroid and should have been an easy target. The team assumed they had failed to detect any echoes because Ceres had a smaller radar cross section than expected.

Following this failure, the Arecibo antenna was fitted with a new transmitter that greatly improved its capabilities. The first asteroid to be observed by the new system was the NEA 1580 Betulia in 1976. Working with Pettengill and Marsden was a graduate student at MIT named Steven Ostro, who did the analysis of the data. He measured Betulia's radar cross section and determined its radius was about 2.9 kilometers. After these successful observations, Ostro began working with Pettengill and Campbell.

They then turned their attention back to the main belt and Ceres. In March and April 1977, weak echoes were finally detected from Ceres. The observations confirmed that Ceres had a lower radar cross section than the Moon, terrestrial planets, or even Eros. They also showed that its surface was rougher than the Moon or planets but smoother than Jupiter's moons, at a scale equal to or greater than the 12.6-centimeter wavelength of the radar signal. This success was followed in November 1979 by detection of weak echoes from 4 Vesta on three nights. In all, six asteroids were observed between 1968 and 1979.[36]

Radar astronomy of asteroids now began an expansion paralleling that of the search programs for near-Earth objects. Between September 1980 and March 1981, another five asteroids were detected by radar: 7 Iris, 16 Psyche, 97 Klotho, 1862 Apollo, and 1915 Quetzalcoatl. The last two were NEAs, and the others were main-belt objects. The round-trip time for the radar signals ranged from 55 seconds for Apollo to nearly half an hour for Psyche. The observations were all made by Steven Ostro, Donald Campbell, and Irwin Shapiro from Arecibo. (Apollo was also independently observed by the Goldstone antenna.)

The echoes showed that Iris and Psyche both had radar reflectivities higher than that of ordinary rocks but below that of pure metal. Klotho, in contrast, reflected the radar signals in a manner like that of ordinary stone.[37] Optical observations, in contrast, indicated Iris was an S-type asteroid, Psyche was M-type, and Klotho was an MP-type.[38] Before the end of 1981 two more asteroids were observed by radar, 8 Flora and 2100 Ra-Shalom, bringing the two-year total to seven, one more than the number observed in the previous eleven years.[39]

It was Steven Ostro who played the major role in this expansion of radar asteroid studies. He had attended Tom Gehrels's third Tucson asteroid conference in March 1979 and caught the asteroid bug. Later in 1979, with his MIT thesis finished, he took a teaching position at Cornell University and began planning a systematic program of asteroid observations. Between 1980 and 1985 he observed 20 main-belt asteroids at Arecibo. By 1992 Ostro had raised his total to 28 NEAs and 36 main-belt objects.

With some five dozen asteroids observed by radar, there was now a statistical basis for understanding the characteristics of asteroid surfaces and composition. This finally settled lingering questions about the composition of 1986 DA. Ostro had observed 1986 DA soon after it was discovered, and its radar brightness indicated that it had at least a metal-rich surface. After five years of analysis and data from other asteroids, it was apparent that 1986 DA was brighter than any of the other asteroids. Ostro and his team concluded that 1986 DA was the first M-type NEA, the remnant of the iron core of an ancient asteroid that had lost its outer crust in a collision.[40]

More remarkable were the indications Ostro was getting about the shapes of NEAs. Starting with his observations of 2201 Oljato during June 12–17, 1983, at Arecibo, he was seeing indications of bimodal echoes. This meant the asteroid was binary—two objects joined together. The data remained ambiguous, however. The echoes strongly indicated binary asteroids, but the modeling and imaging techniques were not adequate to give a definitive answer.[41]

The breakthrough came with the discovery of 1989 PB on August 9, 1989, by Eleanor Helin, setting in motion a chain of events that bordered on the melodramatic. Attempts to observe the asteroid were hampered by a brightening Moon; however, a full lunar eclipse took place on August 16. Astronomers continued to take position data until August 19, when Ostro began his radar observations at Arecibo. As with the lunar eclipse, this was also a matter of luck. Ostro had scheduled time at Arecibo before 1989 PB was discovered, to observe the main-belt asteroid 12 Victoria. By a further coincidence, 1989 PB would be visible only during the same two-and-a-half-hour periods on the same four days that Ostro had scheduled.

Ostro did a few runs on Victoria, which had given a hint of a double-peak echo in 1982, then spent the rest of his observing time on 1989

PB. During the first day of observations, some of the echoes looked strongly bifurcated. It was the first time an asteroid had been resolved as a binary object. Ostro produced 64 images of 1989 PB using the Doppler and range data, each made of about two dozen pixels. Lined up in rows, the images revealed the binary shape and how the asteroid's appearance changed as it rotated. The radar data, combined with other data, also reduced the uncertainty in 1989 PB's position from 15,000 kilometers down to only 300 meters. This resulted in an extremely precise final orbit. Ostro said, "We couldn't lose this asteroid now if we tried." It was subsequently given the name 4769 Castalia.[42]

About a year after the imaging of Castalia, Scott Hudson began working on a modeling technique that would reconstruct an asteroid's three-dimensional shape from the echoes. The result showed Castalia had two equal-sized lobes about three-quarters of a kilometer across, which were joined at a 100- to 150-meter-deep waist. The structure implied that the two lobes had originally been separate objects, with similar compositions and roughness, which had been joined in a low-speed collision.[43]

The next attempt to resolve the surface of an asteroid was during the near approach of 4179 Toutatis in November and December 1992. Ostro put together an eighteen-day observing plan that involved not only the Arecibo antenna but also several of the dishes at Goldstone. The resulting images showed surface details, including impact craters. Toutatis proved to be a contact binary. The two lobes were misshapen objects about 1.6 and 2.5 miles in diameter, each with lumps and ridges. Ostro noted that the resolution was 100 times better than any previous observation of an asteroid. He added, "It's the most complex shape we've seen in the solar system."[44]

The images showed Toutatis as it slowly tumbled through space. Unlike most asteroids, which rotate around a single axis, Toutatis has a dual axis. It rotates every 5.41 days around its long axis, which then precesses around another axis with an average period of 7.35 days. From its surface, the Sun and stars seem to careen slowly through the sky in a manner that never repeats. This behavior is expected for asteroids that have undergone a collision. Most asteroids probably once had this kind of rotation, but over time centrifugal force and internal friction have caused them to settle down to a single rotation aligned with the asteroid's shortest axis.[45]

Ostro's success with Toutatis was repeated during the August 1994 close approach of 1620 Geographos. The radar imagery showed it to be the most elongated asteroid ever seen. Geographos also proved to be slightly larger than expected, with dimensions of 5.1 × 1.8 kilometers. The data indicated that Geographos was a single intact fragment spalled from a larger body following a collision, rather than a collection of loosely bound chunks. The elongated shape had long been suspected from ground observations, as well as laboratory tests indicating that collisions would produce elongated fragments. The extreme shape displayed by Geographos was a rarity, however.[46]

The twin successes of the Toutatis and Geographos observations underlined the capabilities of radar astronomy in the study of asteroids. Radar could resolve the surface features of NEAs. Radar could also provide orbital data more precise than any other means. Radar studies can expand on ground-based data, such as on shapes and mineral content, and radar observations can also be made of a large number of different target asteroids, providing a body of data on the different populations.[47]

THE SPACEWATCH TELESCOPE AND CLOSE CALLS

With the rebirth of asteroid studies in the early 1970s came the idea of a dedicated asteroid telescope. By 1977–1978, the success of the asteroid search program at Palomar Observatory had sparked interest in construction of a specially designed large Schmidt telescope to undertake asteroid surveys. Tom Gehrels disagreed, arguing in 1979 that a large Schmidt would not be effective in this role because any asteroid would trail on a photographic plate, and the trails of the dimmer objects would be too faint to detect. Gehrels suggested that a CCD camera should be combined with automated computer detection software to allow detection of fainter objects while still covering a wide area.

That idea became the Spacewatch telescope, which began observations in May 1983, with regular observations starting on April 22, 1984. The project combined both old and new. While the CCD camera and computer system were cutting-edge technology, the telescope itself dated from 1919. It was a 36-inch Newtonian telescope originally set up on the University of Arizona campus, then moved to Kitt Peak in 1963. Its mirror was glass, rather than the low-expansion materials used in modern telescopes. The telescope would scan three areas of the sky three times each night. A computer then compared the scans, eliminating the fixed stars, to pick out any moving objects. Gehrels's explanation of how it worked brought a big laugh at an asteroid conference in 1983: "We get rid of all the junk—stars and galaxies."[48]

Despite the potential of this automated detection system, the Spacewatch telescope failed to detect any Earth-crossing asteroids during its first five years of operation. Meanwhile, the discovery rate of NEAs was accelerating, and astronomers were realizing that some were coming uncomfortably close to Earth. Between 1982 and 1988 five asteroids and one comet had come within 12 times the Earth-Moon distance. The closest of these passes was 1988 TA, with a separation of 3.5 times the Moon's distance on September 29, 1988. The 1937 close approach of Hermes still held the record, however, at 1.95 times the Moon's distance. But not for long.

Henry H. Holt was the first to notice an asteroid on a pair of films taken on March 31, 1989, with the 18-inch Schmidt at Palomar Observatory. Holt was a volunteer working with Eugene and Carolyn Shoemaker in their asteroid search program. At first the new asteroid, 1989 FC, attracted little attention. When follow-up photos taken on April 2, 3, and 4 were examined, it was noted that 1989 FC's motion against the stars had slowed abruptly. The asteroid was traveling directly away from Earth.

An orbit calculated by Bruce Marsden indicated how close a call it had been. The asteroid was in an orbit with a period of 1.03 years, which stretched from the orbit of Venus out to Mars. On March 22, 1989, it passed Earth at a distance of only 1.79 times the distance of the Moon. It was not detected before the close approach because the Moon was nearly full. Only when 1989 FC moved into the evening sky was it finally spotted. Its exact size was unknown, but if it was an S-type asteroid, the diameter would have been about 220 meters. Had the asteroid struck Earth, it would have blasted a crater 4.7 kilometers across, with a yield equivalent to 430 million tons of TNT. Moreover, it would have struck without warning.[49]

Holt said later, "On the cosmic scale of things, that was a close call."[50] The near-miss of 1989 FC brought to public and political attention the possibility of asteroid impact and set in motion a chain of events. As for 1989 FC itself, it was subsequently named 4581 Asclepius. Others call it by another name: Final Curtain.

It was one of the 13 Earth-crossing asteroids discovered in 1989. The following year saw the discovery of another 14, raising the total known to 85 (the discoveries of the previous two years accounted for 31.7 percent of this total). A part of this increase was due to the first successful detections of Earth-crossing asteroids by the Spacewatch telescope. In late 1989 it was fitted with a new CCD, which had more than 4 million picture elements and a new off-line search routine. On September 25, 1990, the Spacewatch telescope made the first fully automated discovery of an NEA, 1990 SS. In October 1990 the telescope discovered 4 NEAs within an eleven-day period.[51]

The capabilities of the new system were indicated by the number of objects it was finding. Between the first automated discovery on September 25, 1990, and June 30, 1993, Spacewatch discovered 45 new NEAs (more than half the number known through the end of 1990) and recovered another 5 NEAs that had previously been discovered. Each month, some 2,000 main-belt asteroids were also detected. Of those, about 20 percent were previously known. The overall number of NEA discoveries had increased from 7 in 1983 to nearly 40 during 1993.

Because of its large mirror and automated detection system, Spacewatch was able to detect asteroids too faint for other search programs to find. The most significant was 1991 BA, spotted on January 17, 1991, by team member David Rabinowitz, who was later joined by James V. Scotti. It was estimated to be about 30 feet in diameter, the size of a house, and was the smallest asteroid ever discovered. Tracked for five hours, it crossed 7 degrees of the sky. It was heading toward Earth, and only twelve hours later it passed a mere 106,000 miles away, less than half the distance to the Moon. It was the first object known to have passed this close to Earth. Had it hit, the energy released would have equaled about 75,000 tons of TNT, sufficient to destroy a city.[52]

The end of the year brought the discovery of another unusual object. Although 1991 VG's closest approach was 290,000 miles on December 5 (closer than any object up to that time except 1991 BA), its nature remains unclear. Following 1991 VG's discovery on November 6 by Scotti, calculations indicated its orbit was only slightly larger than that of Earth. This very uncharacteristic orbit immediately caused speculation that the object was actually an upper-stage rocket. The results were ambiguous. Light curves from the University of Arizona showed slow variations of a few tenths of a magnitude, consistent with an asteroid with a rotation rate of several hours.[53] On the other hand, when Richard West of the European Southern Observatory observed it through a 61-inch telescope, he reported rapid variation of a magnitude or more, with several bright flashes. He concluded that 1991 VG was an upper stage rotating around more than one axis every six or seven minutes.[54]

There were only a few candidates that could fit the orbit. The S-IVB stages from Apollo missions 8–12 were sent into orbits inside that of Earth. (The S-IVBs from Apollo missions 13–17 impacted the Moon.) This object had an orbit outside that of Earth. Among the possibilities were the upper stages from the Helos 1 and Pioneer 4 launches or the Soviet Luna 1 and 6 flights.

The question of 1991 VG's origin remained unanswered; it could not be found after its close approach on December 5. Attempts by Ostro to get a radar echo in mid-December were unsuccessful. Gehrels believed it was an asteroid, and Marsden was quoted as saying he thought it was an upper stage.[55] Daniel Green of the IAU observed, "The situation is far from conclusive."[56] The now-lost object 1991 VG did acquire a nickname, however. It was called V'ger, after the intelligent alien space probe in the film *Star Trek: The Motion Picture*.

Hermes held the record for closest approach to Earth for more than five decades, but 1991 BA held the distinction for a mere two and a half years. It was displaced by 1993 KA2, discovered by Gehrels with the Spacewatch telescope on May 21, 1993. The asteroid was moving at a rate of 34 degrees per day. A follow-up observation the next day allowed an orbit to be calculated, indicating that on May 20, the day before the discovery, 1993 KA2 had passed within 90,000 miles of Earth and, several hours later, came nearly as close to the Moon. The orbit is highly elliptical, stretching from inside the orbit of Venus out to almost as far away as Jupiter, with a period of 3.3 years.[57]

A third Moon-crossing asteroid was found on March 13, 1994, by Rabinowitz and Scotti with the Spacewatch telescope. The object 1994 ES1 was approaching Earth and would make a close pass on March 15. Although the orbit calculated by Marsden indicated the miss distance would be greater than that of 1993 KA2, he urged other astronomers to track the object as a test to see how quickly the orbit of an object could be refined, if one ever did threaten to hit Earth. The attempt failed because of communications delays with Japan, bad weather in New Zealand, and the inability of an observatory in Australia to locate an object so close to Earth.[58]

Nine months later Scotti and the Spacewatch telescope found a fourth Moon-crossing asteroid. Scotti was observing the comet Brooks 2 on December 9, 1994, when he noticed a fast-moving object. It was 1994 MX1, a bus-sized asteroid. Fourteen hours after it was discovered, it passed within 65,000 miles above the northern hemisphere. Because of its small size, Scotti said, "it probably wouldn't have survived coming through the atmosphere. We probably would have seen a great fireball." He also pointed out a more ominous factor. "While I did find this little guy, I missed the other 40 or 50 objects of similar size that probably also passed within the Moon's distance that day."[59]

5

FAR FRONTIERS: FROM THE TROJANS TO THE KUIPER BELT

Reaching to some great world in ungauged darkness hid.

Coventry Patmore, *The Unknown Eros,* 1877

WHILE MOST ASTEROIDS ORBIT in the main belt between Mars and Jupiter, others exist in realms farther afield. They are on the far edge of the main belt, sharing the orbit of Jupiter, and between the orbits of Jupiter and Saturn. In recent years, a separate asteroid belt was discovered in the darkness at the outer rim of the solar system. These objects are so far away that the Sun is only the brightest star, and their ice is frozen as hard as iron.

ULTIMA THULE

Beyond the rim of the asteroid belt are the Hilda asteroids. As with the near-Earth asteroids, the group is defined by their orbits. The first was 153 Hilda, discovered by Johann Palisa on November 2, 1875, and named for the eldest daughter of the Austrian astronomer Theodor von Oppolzer. By the late 1980s 31 numbered Hilda asteroids had been found, and the number grew to 56 in the early 1990s. They are at about 4.0 AU from the

Sun (the main belt is between 2.0 AU to 3.5 AU). Still farther out, at 4.26 AU, is the asteroid 279 Thule. It was also discovered by Palisa, on October 25, 1888. The name refers to the island that marked the northern limit of the habitable world for the Romans—Ultima Thule.

Both the Hilda asteroids and Thule break a pattern that holds for the inner asteroid belt. The Hilda asteroids have orbits at the 3:2 resonance point with Jupiter, while Thule orbits close to the 4:3 resonance point. In other cases, the resonance points have been swept clear of asteroids, forming the Kirkwood gaps. The reason for this exception seems to be that although the orbits of the Hilda asteroids are close to the orbit of Jupiter, the objects themselves do not make close approaches. The Hilda asteroids have fairly eccentric orbits, ranging between 3.4 AU and 4.6 AU. When the objects are close to Jupiter's orbit, Jupiter is always far away. As Jupiter approaches their orbits, the asteroids themselves are heading for their perihelion points. This elegant balancing act keeps them out of the gravitational clutches of Jupiter. In the case of Thule, its orbit is not very eccentric, but it never comes closer to Jupiter than 1.1 AU. This also protects Thule from Jupiter's gravitational effects.[1]

As with the near-Earth asteroids, there is speculation that the Hilda asteroids were originally comets. Some comets captured by Jupiter have orbits similar to those of Hilda asteroids. A comet in a low-eccentricity orbit between Jupiter and Saturn can be sent into an orbit with a perihelion well inside the orbit of Jupiter. The data on the composition of Hilda asteroids is very limited. Only P-, C-, and D-types have been identified among them; there were no M- and S-types. As this is typical of outer-belt asteroids, it is not possible to prove or eliminate comets as a source.[2]

THE TROJAN ASTEROIDS

On February 22, 1906, Max Wolf photographed a dim asteroid that was subsequently designated 1906 TG. When an orbit was calculated for the object, it was nearly circular at the distance of Jupiter. It was not clear how the asteroid could keep this position in the face of Jupiter's gravitational attraction. It was C. V. Charlier of the Lund Observatory in Sweden who realized that 1906 TG was the first practical demonstration of what had been only a mathematical curiosity.

In 1772 the French mathematician Joseph Louis Lagrange had shown that it was possible for two bodies in independent orbits around the Sun to maintain a constant position relative to each other. This occurred because the gravitational attractions of the Sun and planet on the smaller third object balance out. Three of these Lagrangian points are on a line through the Sun and planet: the L-1 point is between the Sun and the planet, L-2 is beyond the planet opposite the Sun, and L-3 is in the same orbit as the planet but remains directly opposite it on the other side of the Sun. These three positions are only marginally stable, however. If the small object is displaced at a right angle to the Sun-planet axis, it will return to the original position. If the displacement occurs along the axis, then the object will drift away and never return. Two other Lagrange points, however, are stable. The L-4 point is ahead of the planet, and the L-5 point trails behind it. The L-4 and L-5 points are an equal distance from the Sun and planet. An object at the L-4 and L-5 points can be displaced in any direction and it will still return.[3]

The asteroid 1906 TG was 55.5 degrees ahead of Jupiter—the L-4 point is 60 degrees ahead of the planet. The asteroid was subsequently named 588 Achilles, after the Greek hero of the Trojan War who was the central figure of Homer's Iliad. The name was proposed by Palisa, who later named two other asteroids at Lagrangian points. The asteroid 617 Patroclus was named for a Greek warrior who was a friend of Achilles, and 624 Hektor was named for the Trojan champion who killed Patroclus and was, in turn, killed by Achilles. Although Hektor (a Trojan) was at the L-4 point with Achilles (a Greek), and Patroclus (also a Greek) was at the L-5 point, a pattern developed of naming asteroids at the L-4 point after Greek warriors, and those at the L-5 point were given Trojan names. For this reason, this class of asteroids became known as Trojan asteroids.[4]

As with the near-Earth asteroids, discoveries of new Trojans were sporadic. The first three discoveries in 1906 and 1907 were followed by 659 Nestor by Max Wolf on March 23, 1908. It was nine years before he found the next Trojan, 884 Priamus on September 22, 1917. For the next two decades all Trojan discoveries belonged to Karl Reinmuth. His first was 911 Agamemnon on March 19, 1919. Eleven years would pass before he found another. In 1930 Reinmuth found three Trojans (two on October 17), and one each in 1931, 1936, and 1937. World War II ended the discovery of Trojans, and it was not until September 23, 1949, that Reinmuth found his next, 1749 Telamon.

Discoveries continued to be made during the succeeding decades. The Palomar-Leiden Survey found 15 Trojan asteroids in 1960, and another survey in 1973 found 34. Eugene Shoemaker undertook a systematic search for Trojans in 1985 and a second effort in 1988. Together, they added another 45 Trojan asteroids. All told, some 300 Trojan asteroids had been found by the early 1990s. The Trojans given permanent numbers were more limited; there were 46 numbered Trojans in mid-1987 and 170 by late 1999.

The orbital behavior of the Trojan asteroids is very complex, in contrast to Lagrange's idealized solution, where the objects are in circular orbits and unaffected by any other bodies in space. In reality, the orbits are elliptical with differing orbital inclinations, and Saturn's gravitational attraction also has an influence. As a result, the Trojans are not actually at the L-4 and L-5 points but rather trace out narrow looping paths around the Lagrangian points. As an asteroid moves away from Jupiter, the planet's gravitational attraction changes in strength and direction, slowing the asteroid down. At the far end of the loop, their speeds are equal. As they continue around the Sun, the asteroid is traveling more slowly than Jupiter, and the distance between them starts to close. The maximum difference in speed occurs when the two are 60 degrees apart (i.e., at the Lagrangian point). At the near end of the loop, their speeds are again identical. At this point, the asteroid's relative velocity becomes greater, and it begins the cycle again.[5]

The Trojans can wander far afield of the Lagrangian points, varying from as close as 45 degrees from Jupiter, out to 80 or 100 degrees away. Each cycle is a slow process, taking between 150 to 200 years to complete. This behavior even led to the naming of 1868 Thersites. Thersites was a Greek warrior who wanted to end the siege of Troy and go home. It was a fitting name for the Trojan asteroid found farthest from the Lagrangian point.

The shape of Hektor, which is the largest Trojan, has also been a subject of considerable speculation. Its light curve would indicate a cylinder 300 kilometers long and 150 kilometers wide, but how so large an object could have been formed in such a nonspherical shape is not clear. No less than four shapes have been proposed for Hektor: two different elliptical shapes, a dumbbell shape formed by two nearly spherical objects pressed together, and two elongated objects nearly touching. Unlike asteroids in the main belt, the Trojans drift around the Lagrangian points with low relative velocities. Rather than shattering each other, it is possible that they might stick together or go into orbit around each other.[6]

The two current theories of the origins of the Trojan asteroids are that they formed at the Lagrangian points or that they were captured by Jupiter's gravity in the past. Related to this is the origin of Jupiter's outer moons. Kuiper suggested in 1956 that Jupiter's moons may have been Trojans captured when they drifted too close to the planet. Conversely, the Trojans may be moons that escaped from Jupiter.

The composition of both bodies gives some indications. About 40 percent of the Hilda asteroids are P- and D-type asteroids, but these two types account for about 70 percent of the Trojans. In contrast, the outer moons of Jupiter seem to be similar to C-type asteroids. The Trojan asteroids also seem to be different from the main-belt objects but similar to objects found in the outer solar system. This implies that Trojans and outer-belt objects are related.

There do not seem to be any differences in composition between the objects at the two Lagrangian points, but there does seem to be a difference in the numbers. According to one estimate from the mid-1980s, the Greeks at L-4 outnumber the Trojans at L-5 by 700 to 200. This is thought to be the result of the gravitational effects of Saturn. Because of this, the Trojans are not as stable as theory would indicate. They can be dislodged from their gravitationally protected realm and sent to roam the solar system. It has been suggested that some of the Jupiter-group comets are actually Trojan asteroids that escaped in this manner. It is known that comet Slaughter-Burnham was captured into an orbit with temporary Trojan-like behavior, but it is not clear that the reverse could also occur.[7]

In the early 1990s Shoemaker proposed that the Trojans are leftover samples from the formation of Jupiter, similar in composition to comets. In his model, 4.6 billion years ago there were planetesimals with a total mass about 60 times that of Earth in the region currently between the orbits of Jupiter and Saturn. As Jupiter moved inward, it swept up about 10 Earth masses of material to form its rocky core. The rest of the planetesimals were flung out of the solar system. A small percentage, however, underwent repeated close encounters with Jupiter and eventually became trapped at the Lagrangian points. The benign conditions at the Lagrangian points, with few collisions between the asteroids, have preserved them in nearly pristine condition ever since. Shoemaker estimated the total number of Trojan asteroids 15 kilometers or larger at about 2,300.

A related question is the existence of Trojan asteroids around other planets. Between 1978 and 1982 Scott Dunbar searched for Earth Trojans with the 48-inch Schmidt on Palomar Mountain. A parallel effort was done by Francisco Valdes and Robert Freitas with the 24-inch Schmidt at Kitt Peak in 1981 and 1982. Each time, nothing was found.

It was nearly another decade before the existence of a non-Jupiter Trojan was confirmed. On the night of June 22, 1990, Henry Holt and David Levy were taking a series of photos with the 18-inch Schmidt to search for near-Earth asteroids. The next day, Levy noticed an unusual asteroid trail on one of the pairs of photos. It was longer than and at a different angle from most asteroids. They rephotographed the area the next night and confirmed its existence.

The first to realize the significance of 1990 MB was Edward Bowell of Lowell Observatory. He immediately identified it as the first Martian Trojan. Brian Marsden, the director of the Minor Planet Center, initially did not believe it. Later he said the discovery was "a bit of a surprise." To convey the sense of elation at the discovery, the first Mars Trojan was given the name 5261 Eureka.

Eureka itself is about 2 kilometers in diameter and is in a stable orbit. Its orbital behavior is similar to that of the Jupiter Trojans. From the L-5 point, it covers a loop that stretches along 80 degrees of the orbit of Mars. A cycle takes about 1,000 years to complete. As with the Jupiter Trojans, Eureka's origin is a question mark. Current computers are unable to make long-term simulations of the orbital evolution of the inner planets. Thus, it is not yet possible to determine how long it has been at the Martian Lagrangian point. Shoemaker believed that Eureka is too small to have survived for 4.5 billion years in its current orbit. He concluded that it was captured sometime within the past 500 million years.

Eureka also raises the question of whether other planets have Trojan asteroids as well. According to some calculations, the stable regions around Saturn, Uranus, and Neptune are so small that it is unlikely that they would have any such objects. Others feel that it would be worth looking for Saturn Trojans, although they would be small and difficult to spot, and the effects of Jupiter's gravity would make it uncertain where to look.[8] Despite the repeated failure to find Earth Trojans, such objects could theoretically exist. Paul R. Weissman and G. W. Wetherill of the

University of California calculated in 1974 that despite the gravitational effects of Venus and Jupiter, an object could be stable at Earth's L-4 and L-5 points for at least 10,000 years and possibly longer.[9] Given the complex loops around the Lagrangian points such objects make, and their probable small size, discovering them would be difficult. Since Eureka's discovery, it has remained "a real unicorn in the astronomical zoo."[10]

2060 CHIRON

The first indication that there might be asteroidlike objects in the outer solar system came on October 31, 1920, with the discovery of 944 Hidalgo by Walter Baade. Hidalgo has an orbit unlike any known up to that point. It stretches from 2.0 AU, at the inner edge of the asteroid belt, out to 9.7 AU, which is just beyond the orbit of Saturn. The orbit is also highly inclined, at 42.4 degrees, and has a period of 14.2 years. This orbit would imply that Hidalgo is a burned-out comet, but it is 40 to 60 kilometers in diameter, which is considerably larger than the typical comet nucleus. It appears to be a D-type asteroid, but color changes have been detected in the course of its ten-hour rotation. No trace of cometary activity has ever been detected in Hidalgo since its discovery. Besides the possibility that Hidalgo is a comet nucleus, it has also been proposed that it is an escaped Trojan or that it originated in the outer asteroid belt and was subsequently sent into its present orbit. Yet, any process that would have placed it into its current orbit should have also acted on other objects, but for 57 years no other asteroid was known to travel beyond Jupiter.

During the late 1970s Charles T. Kowal was undertaking a search for unusual objects in the solar system. Each month he would use the 48-inch Schmidt on Palomar Mountain to photograph the night sky directly opposite the Sun. Called the opposition point, there any asteroids would be moving westward at a rate proportional to their distance from Earth. A nearby Apollo asteroid would leave a long trail on the plate, while a distant object would move only slightly over the course of a day. Kowal would look for any objects moving very fast or very slowly. On October 18, 1977, Kowal photographed a slowly moving object. The object's rate of motion indicated it was at the distance of Uranus. The planet and its satellites, however, were in the opposite side of the sky, and nothing else was known to orbit at that distance.

Such a unique object required an equally unique name. Kowal found that no asteroid had been named after the mythological Centaurs. The wisest and most just of this unruly breed was Chiron. He was the son of Saturn and the grandson of Uranus, which Kowal thought tied in with the object's orbit. So 1977 UB was named 2060 Chiron. Kowal also suggested than any further objects found be named after the Centaurs and that they be called the Centaurian asteroids.[11]

Photos taken in October and November 1977, along with two images taken at Palomar in September 1969, allowed a preliminary orbit for Chiron to be calculated. With this, a search for prediscovery images could be made, in order to refine the orbit. Images were found from 1895, 1941, 1943, 1945, 1952, and 1976, indicating that Chiron currently has an orbit of 50.7 years, which ranges from nearly the orbit of Uranus to just inside the orbit of Saturn.[12] The orbit was chaotic, however, meaning each revolution would be different because of the gravitational attractions of the outer planets. Brian Marsden calculated backward and found that in 1665 B.C. Chiron passed within 0.1 AU of Saturn (about the distance of Saturn's moon Phoebe). As a result, it is not possible to calculate Chiron's orbit before then. Looking ahead, to A.D. 7400, Chiron's orbital period will be less than 46 years, while its inclination will increase from the current 6.9 degrees to more than 9 degrees. Chiron's life expectancy is very brief on the cosmic scale. In about a million years it will either collide with a planet or be ejected from the solar system.[13]

"It is curious how people need to label things, and to place everything into pigeonholes," Kowal noted. "Chiron," he observed, "simply does not fit into any pigeonhole." Based on Chiron's orbit, it is tempting to call it a comet, but it is about ten times larger than the typical comet nucleus. Although initial press reports referred to it as a possible tenth planet, Chiron is also about ten times smaller than the smallest planet. Although Chiron resembles a C-type asteroid, Kowal felt that Chiron should not be considered an asteroid, as it is probably different in composition and origin. In a letter published in the March 1978 issue of *Sky and Telescope* magazine, Kowal concluded, "for now Chiron is just—Chiron."

In that same issue, a letter from George Wallerstein of the University of Washington was published, making what was to be a prophetic observation: "In some ways, it is more similar to Pluto than to other asteroids, suggesting that there is a large class of such objects of which Pluto is the largest and Chiron the nearest. Hence, a vast number of asteroids may await discovery outside the orbit of Neptune."[14]

Over the following years, Chiron's orbit took it closer to the Sun. In February 1988 David J. Tholen of the University of Hawaii and his colleagues observed Chiron to be 0.7 magnitude brighter than calculated. Why was not immediately clear. The observations did not indicate a cometlike outburst, as a CCD image taken on February 21 did not show a coma, and a spectrum failed to indicate any emission features.[15] The observations were confirmed on March 23 by Schelte J. Bus and two other astronomers at Lowell Observatory. They found Chiron was 0.6 magnitude brighter than it had been in 1986, when a series of observations were made to record Chiron's light curve. The new measurements matched the previous curve but were brighter.[16]

The brightening continued, and on April 9, 1989, a coma was detected around Chiron by Karen J. Meech and Michael Belton. They used the 4-meter telescope on Kitt Peak and confirmed the discovery the following night.[17] At a distance from the Sun of some 13 AU, Chiron is too cold for water ice to sublimate off into space. Instead, the coma is formed from sublimating carbon monoxide, methane, or nitrogen ice. In 1990 cyanogen gas was detected in Chiron's coma. Although this is only a trace substance in comet comas, it can be more easily detected than other gases because it fluoresces in sunlight.[18]

Outbursts of comets at such great distances from the Sun were previously known. A coma has been detected around comet Bowell at a distance of 13.6 AU, while the brightness of comet Schwassmann-Wachmann 1 has sporadically increased by a factor of 100, even though it is never closer to the Sun than 5.5 AU.[19] What came as a surprise was the level of activity shown by Chiron. Over periods of weeks or months Chiron can vary in brightness by between 30 and 50 percent. There have also been random changes of a few percent detected from one night to the next and even on an hour-by-hour basis. Another surprise came when prediscovery photos of Chiron were examined. They showed that Chiron was actually brighter between 1969 and 1972, at a distance of 19.5 AU, than it was in the early 1990s, as it neared its closest approach to the Sun. The process was active even under those far colder conditions.[20]

Added information came from a near-occultation of Chiron on March 9, 1994. It was observed from South Africa and from aboard the Kuiper Airborne Observatory flying off Brazil. Although Chiron just missed the star at both sites, four drops in brightness were observed. The sharpest, labeled F1, is probably caused by a dense, narrow dust geyser on Chiron. Another dimming, F2, is believed to be caused by a second geyser. Superimposed on these is a much broader, slight dimming called F3, thought to be caused by a large, spherical dust cloud held by Chiron's gravity. A final event, F4, is thought to be a kind of tail: dust and gas pushed directly away from the object by light pressure and the solar wind.[21]

Over the past two decades a clearer picture of Chiron's nature has become apparent. Based on infrared measurements, it is 170 kilometers in diameter, with an error of plus or minus 20 kilometers. It is nearly spherical in shape and has a dark surface crust of silica dust or organic compounds. Beneath that are exotic, low-temperature ices, which erupt from fissures in the form of geysers throughout its orbit. Because of the low-temperature nature of these ices, Chiron cannot have been in its current orbit since the formation of the solar system, nor could it have been closer to the Sun. In either case, the ices would be long gone by now. This confirms calculations that show Chiron's orbit is unstable over periods of a few million years. But the most basic discovery was that Chiron was not alone.

THE KUIPER BELT

Like so much in the history of the asteroids, the idea of an asteroid belt outside the orbit of Neptune long predated its discovery. In 1951 a book titled *Astrophysics* was published—a mid-century survey of astronomy and where it might be headed. One of the contributors was Gerard Kuiper, who wrote an overview of solar system astronomy. He discussed an unusual situation in the outer solar system: there seemed to be a boundary at 30 AU from the Sun. The solar system just seemed to stop abruptly. Kuiper could not see a reason for this, as the original solar nebula, from which the planets had condensed, would have extended out farther. Kuiper concluded that there were a large number of small icy bodies on the rim of the solar system. They had not formed a large planet because they were too spread out to collide.[22]

Kuiper was not alone in his speculation. In 1943 and 1949 the Irish astronomical scholar Kenneth Essex Edgeworth had wondered about this seeming cutoff and also proposed that there might be small bodies in the darkness. The suggestions of both men were ignored. In part this was because solar system astronomy was an outcast (it has been said that Kuiper was two-thirds of the entire planetary astronomy community in the 1940s), but another factor was the lack of technology to detect such small, dim bodies.[23]

Not until the early 1980s was the idea revived, based on both observational data and theoretical calculations. In 1983 the infrared astronomy satellite discovered dust disks around several nearby stars, which were interpreted as the result of dust from asteroids and comets in orbit around the stars. During the 1980s theoretical calculations indicated that short-period comets (that is, comets with orbits of 200 years or less) had to come from a region outside the orbit of Neptune. Both these events sparked a revival of interest in the Kuiper Belt.[24]

A search for such objects would not be easy. Clyde W. Tombaugh had made a systematic search for a planet beyond Neptune between 1929 and 1943 and found only Pluto. Kowal's later search could detect objects 100 times fainter than Pluto, but only Chiron was spotted. Given the distance, it would be possible only to find larger bodies of several hundred kilometers in diameter. They would still be as dim as magnitude 24, 10,000 times fainter than Pluto. Several groups searched unsuccessfully for Kuiper Belt objects in the late 1980s.

Additional asteroids were found in the outer solar system, however. On February 18, 1991, Robert H. McNaught spotted a 5-kilometer asteroid in a highly elongated orbit, which stretched from inside the orbit of Mars out to a distance of 22.2 AU, well beyond the orbits of Uranus and Chiron. The orbit is inclined 62 degrees and has a period of 41 years. At the time of its discovery, 1991 DA was just crossing the orbit of Mars and heading back out into space. Because of this unusual orbit, it was thought that the object might be a comet. Subsequent observations showed no signs of a coma or cometary emissions, however. It did brighten by half a magnitude in only 40 minutes, indicating a possible rapid rotation. The exact nature of 1991 DA remains unclear. It could be a comet that has lost its gas and dust or an ordinary asteroid that has somehow been sent into a highly inclined and elongated orbit.[25]

Less than a year later, on January 9, 1992, David L. Rabinowitz discovered 1992 AD with the Spacewatch telescope. After three weeks of observations, an orbit could be calculated. It showed 1992 AD was the outermost asteroid then known. Its orbit ranges from 8.7 AU, just inside the orbit of Saturn, to 32 AU, well outside the orbit of Neptune. At its farthest point from the Sun, it is 2 billion kilometers more distant than Chiron. Its orbital period is 93 years; Chiron's is 51 years. Initial size estimates put its diameter at about 160 kilometers, making it smaller than Chiron.[26] The asteroid was named 5145 Pholus.

Observations showed that Pholus was redder than any asteroid or comet previously observed. Several astronomers suggested that this was the result of a thin layer of methane frost being subjected to ultraviolet light or particle radiation, converting the material to a complex organic material. This indicated that Pholus was a primitive body unchanged by a close approach to the Sun. No indications of a coma were seen.

Like Chiron, Pholus was also in a chaotic orbit influenced by Saturn. Pholus was estimated to have a 50 percent chance of being ejected from the solar system in a period of 2 million years. The odds of its orbit evolving into that of a Jupiter-family comet were about 40 percent, while there was a 1 to 2 percent chance that it would be sent into an Earth-crossing orbit over a million-year period.[27]

Among those searching for Kuiper Belt objects were David Jewitt and Jane X. Luu. Both had made long journeys. Jewitt was a British expatriate astronomer on the staff of the University of Hawaii, and Luu was a Vietnamese refugee who had, as a child, escaped just before the fall of Saigon to the Communists. For the search, they used the 2.2-meter telescope on Mauna Kea. This observatory is the stuff of astronomical legend. Built atop an extinct volcano, the steep road is not for the timid. The observatory itself is at 13,796 feet, and astronomers often suffer from lack of oxygen. One tale is told of an astronomer who forgot to turn out the darkroom lights when he developed a photographic plate.

Jewitt and Luu began their search in 1987, but for five years nothing was detected. None of the other groups found anything either. It was clear that any Kuiper Belt objects were very faint. More sky would have to be searched, and to fainter magnitudes. Several of the search groups dropped out of the chase, but Jewitt and Luu continued.

Finally, on August 30, 1992, they spotted a tiny, dim object amid the star fields on the CCD images. From its motion, Jewitt and Luu could tell it was beyond Neptune. To eliminate any possible mistakes, they followed it for several nights until the Moon made observations impossible. The object was designated 1992 QB1, but its nature was still uncertain.

Some of the calculated orbits indicated it was a Kuiper Belt object, but the orbital arc was so short it was still possible it was on an elliptical orbit that passed inside the orbit of Neptune. They continued to observe 1992 QB1 for several months.[28] Their observations showed that it was in a near-circular orbit from 39.6 AU to 49.1 AU, well outside the orbit of Neptune. The object is also between 200 and 240 kilometers in diameter and reddish in color. Its orbital period is 296 years. It was, in fact, the first object found in the Kuiper Belt.

Even as Jewitt and Luu observed 1992 QB1, they continued searching, and on March 28, 1993, they found 1993 FW. It was a near-twin to 1992 QB1; both were about 200 kilometers in diameter and reddish. But 1993 FW had a more distant orbit, ranging from about 38 AU out to 56 AU, with a period of some 322 years. Also, its orbit was inclined 7.7 degrees, more than the 2.2 degrees of 1992 QB1. Both were thousands of times fainter than Pluto and more than 50 times fainter than any object Kowal's search could have located.

Jewitt and Luu had already picked unofficial nicknames for the two objects. They were fans of the spy novels of John La Carre, and so 1992 QB1 was nicknamed Smiley, after the British spy in several of the books, and 1993 FW became Karla, for the Soviet master spy who was his nemesis.[29]

Smiley and Karla did not remain out in the cold for very long. On the nights of September 14 and 17, 1993, Jewitt and Luu found 1993 RO and 1993 RP. On September 22 another two Kuiper Belt objects were announced, 1993 SB and 1993 SC, found by Iwan P. Williams, Alan Fitzsimmons, and Donal O'Ceallaigh using the 2.5-meter Isaac Newton telescope at La Palma in the Canary Islands.[30]

All four of the new objects had similar elliptical orbits, rather than the circular orbits of 1992 QB1 and 1992 FW. All four passed close to the orbit of Neptune, but without crossing it, rather like the orbit of Pluto, which is elliptical but does cross that of Neptune. The two planets have a stable resonance: Neptune completes three orbits for every two made by Pluto. Preliminary orbits for the four new Kuiper Belt objects also indicated a 2:3 resonance, which prevents them from coming any closer to Neptune than 13 to 15 AU and keeps them out of the planet's gravitational influence.[31]

By September 1996 the total had reached 39. Jewitt and Luu found 9 more Kuiper Belt objects by late 1997 using a new large CCD array on the 2.2-meter telescope, which allowed them to cover a greater area in each image. The total was again raised, to 56 (including Pluto). By the end of 1997 the total stood at 61.[32]

The objects showed individual peculiarities. The three discovered in September 1996 all orbited at the distance of Pluto and were all between 150 and 200 kilometers in size. In contrast, one of Jewitt and Luu's discoveries, 1996 TO66, was some 800 kilometers in diameter, almost one-third the diameter of Pluto. Still another of their discoveries had an extraordinary orbit. The distance of 1996 TL66's closest approach to the Sun is 35 AU, almost the same as Neptune. Unlike other Kuiper Belt objects, which are in circular or slightly elliptical orbits, 1996 TL66's farthest point from the Sun is 132 AU away. This is equivalent to more than 19 billion kilometers, considerably farther than the boundaries of the previously discovered Kuiper Belt objects, which were between 30 and 50 AU. It takes 1996 TL66 nearly 800 years to complete a single orbit. Its elongated orbit was believed to be the result of an encounter with Neptune or a large object within the Kuiper Belt.[33]

While the number of Kuiper Belt objects increased, so did the population of Centaurian asteroids. On April 26, 1993, the Spacewatch telescope detected 1993 HA2, an object that proved to be even redder than Pholus. The same telescope also spotted 1997 CU26, which, based on its brightness, was slightly larger than Chiron, which had been the largest Centaurian asteroid. It also brought the total of known Centaurian asteroids to seven.[34]

There were more surprises awaiting in the darkness. When 1996 PW was first spotted by Eleanor Helin on August 9, 1996, it was at a distance of 2.5 AU, between the orbits of Mars and Jupiter and in the heart of the main belt. The object itself was estimated to be between 8 and 15 kilometers in diameter. It seemed to be just another small asteroid, but when an orbit was calculated, it was apparent how unusual 1996 PW was. The far point of its orbit was 645 AU from the Sun, nearly 100 billion kilometers. The object took some 5,800 years to complete an orbit. This placed it in the Oort Cloud, which is the source of long-period comets. There are many billions of comets orbiting at this vast distance, where they remain until a few have their orbits disturbed by the gravity of a passing star and then begin the long fall toward the Sun. No object that was not a comet had ever been seen in this type of orbit. Yet 1996 PW showed no sign of a coma or tail. Except for its orbit, it was an ordinary asteroid.

The possibility that there might be asteroids as well as comets in the

Oort Cloud was first suggested by John A. Wood, of the Harvard-Smithsonian Center for Astrophysics, in 1979. Following the discovery of 1996 PW, his idea was expanded on by Paul R. Weissman, of JPL, and Harold F. Levison, of the Southwest Research Institute. According to their calculations, 1996 PW had its origins in the disruption of the early asteroid belt. During this period, several trillion main-belt asteroids could have been flung out of the inner solar system by the gravitational effects of Jupiter. Some ended up in the Oort Cloud, amounting to at least 1 percent of its population. Like comets, a few would then be sent back toward the Sun.[35]

Ground-based telescopes are not the only ones looking for Kuiper Belt objects. The Hubble space telescope was used to estimate the number of small objects. Anita Cochran, of the University of Texas, and her colleagues combined 34 images of a single field, removed the stars and galaxies, then combined them again, looking for dim moving objects. Rather than the large, 200-kilometer objects found by ground-based searches, Cochran was looking for objects the size of large comet nuclei, about 20 kilometers in diameter. Several hundred faint blips were found near the Hubble's magnitude 28 limit. Most were electronic noise, but several dozen appeared to be real objects. This would indicate that there were several thousand of these small objects in every square degree along the ecliptic, supporting the idea that the Kuiper Belt was the source of short-period comets.[36]

ORIGINS OF THE KUIPER BELT

Before the discovery of the Kuiper Belt, the outer solar system did not make sense. Pluto had been a curiosity ever since its discovery. Its elliptical, inclined, and overlapping orbit was immediately recognized as different from the neat, nearly circular bull's-eye orbits of the other planets. Lowell's original calculations had indicated a Planet X with a mass between that of Neptune and Earth. When Pluto was discovered, it was variously estimated to have a mass as great as 0.8 or 0.9 of Earth or as small as 0.1. As the decades passed and more observations were made, the estimates for Pluto kept getting smaller. Following the discovery of Pluto's moon Charon, a value of 0.0023 the mass of Earth was calculated for Pluto, and Hubble images of Pluto showed it was smaller than Earth's Moon.[37]

Nor were the troubling oddities limited to Pluto's existence. Neptune's moon Triton orbits in the reverse direction from the planet's other moons. Additionally, Uranus tilts 97.9 degrees. In effect, it orbits the Sun on its side. Neptune's axis is tilted 29.6 degrees, and Pluto tilts 122 degrees. The ratio between the mass of Pluto and its moon Charon is greater even than that of the Earth and Moon, which had been the sole example of a double planet.

The discovery of the Kuiper Belt provided proof for ideas that had been developing over the previous years about the origins of the outer solar system. Collisions had a major role in the early history of the solar system. The Apollo data indicated that the Moon had been formed in the collision of the proto-Earth and a Mars-sized planetesimal. The collision blasted a large amount of the proto-Earth's mantle into orbit, which, within a day, formed the proto-Moon. In the mid-1980s scientists began to realize that a similar event was responsible for the formation of Pluto's moon Charon. Pluto was hit by an object one-tenth to one-half its size. The impact sent debris (which became Charon) into orbit around Pluto and also resulted in the extreme tilt of Pluto's axis.

Triton's unusual orbit was the result of a near-collision with Neptune. By some process, it was slowed enough for it to be captured by Neptune. Triton's reversed orbit was a prime indication that such a capture event had occurred. The data on Triton from the Voyager spacecraft, as well as ground-based observations of Pluto, also provided a major clue. They showed that the two objects were nearly identical in composition and were totally unlike any other solar system objects, indicating they had similar origins.

The odds that Pluto would collide with that object and that Triton would be captured are vanishingly small, however, even over the 4.5 billion years of the solar system. If there were only two bodies (Pluto and the object that hit it), it would take between 10,000 and 10 million times the age of the solar system for them to collide. To get around this, statistical calculations indicated there would have to be between 300 and 3,000 Pluto-like objects in existence during the early days of the solar system, increasing the odds of such unlikely events. It also meant that Pluto and Triton were not oddball little worlds but rather survivors of a huge number of similar bodies.

The tilts of Uranus and Neptune also provided evidence for both massive impacts and a large number of objects. Calculations indicate that creating their tilts would require the impacts of Earth-sized or larger bodies, and there would have to be 10 to 100 objects with mass a quarter or greater than that of Earth within the solar system. The massive size of Jupiter and, to a lesser extent, Saturn protected them from such tilts.

A picture began to become clear. As Uranus and Neptune grew, their effect was similar to Jupiter's on the early main-belt asteroids: the small bodies at the edge of the solar system, such as Pluto and Triton, were prevented from growing any larger. Rather, the objects were stirred up, causing them to collide and shatter. The large objects inside the orbits of the outer planets were quickly swept up or thrown out of the solar system. Objects just outside the orbit of Neptune also began to be cleared out. Some were thrown farther out, while some went into chaotic orbits between the outer planets, much like Chiron. Pluto remained because of its 2:3 resonance with Neptune. Triton survived because it was captured by Neptune. Both objects were protected from the gravitational effects that removed the other objects.

These theoretical ideas had come together by the spring of 1991. They explained the strange nature of the outer solar system. The problem was that all the evidence was neatly removed several billion years ago. It was seen by some astronomers as a nice theory, but one lacking in proof. A little more than a year later, 1992 QB1 was found, to be followed by the host of other Kuiper Belt objects, the remnants of the objects that had populated the outer solar system. Originally they amounted to 10 to 50 Earth masses. Today, only one-quarter to one-half an Earth mass remains.[38]

As with the main-belt asteroids, there seem to be different types of Kuiper Belt objects. When Stephen C. Tegler of Northern Arizona University and William Romanishin of the University of Oklahoma surveyed sixteen Kuiper Belt objects and Centaurian asteroids, they were surprised to find two distinct types. One type showed a neutral spectrum, similar to main-belt asteroids, and the other type included the reddest objects in the solar system. Another surprise was that the two types did not have distinctive orbital characteristics. Chiron and Pholus have similar perihelion distances, but the first has a neutral spectrum and the other is red. The two types seem to be the result of their size, composition, impacts, or some unknown factor.[39]

There also seem to be Kuiper Belt families. In October 1999 Alan Stern, Robin Canup, and Daniel Durda, astronomers with the Southwest Research Institute, suggested that several Kuiper Belt objects were debris created by the Pluto impact. This was based on several types of evidence. Their close orbital similarities were consistent with the expected debris distribution from the collision. The color of Pluto is similar to some Kuiper Belt objects, and the color of Charon is like that of other, different objects, which implied similar surface compositions. Finally, the sizes of the objects are similar to theoretical calculations of collisions and those seen in main-belt asteroid families.

Although the astronomers noted that further research was needed, there were several important implications. First was that the precesses that created main-belt families also operated in the Kuiper Belt. Another was the suggestion that a small fraction of comets may actually be samples of Pluto and Charon.[40]

6

ASTEROID SPACE MISSIONS

Let us create vessels and sails adjusted to the heavenly ether, and there will be plenty of people unafraid of the empty wastes.

Johannes Kepler, letter to Galileo, April 1610

ON OCTOBER 4, 1957, the USSR launched the first satellite, Sputnik 1, beginning the second great age of exploration. In 1962 the U.S. Mariner 2 space probe made the first successful flyby of Venus, and in 1965 Mariner 4 radioed back the first photos of Mars.[1] No longer would astronomers have to study asteroids from Earth. The possibility now opened of making flyby missions to photograph their surfaces, of orbiting them to conduct detailed mapping and mineralogic surveys, and, finally, of collecting samples to return to Earth.

PIONEER 10 AND 11

Pioneer 10 and 11 were the first two spacecraft to cross through the asteroid belt. They were simple, low-cost, spin-stabilized probes with three mission goals: explore the interplanetary medium beyond Mars, assess the dangers posed by debris in the asteroid belt, and explore Jupiter. One unknown in planning the missions was the number of meteoroids in the asteroid belt. The fear was that over billions of years the collisions of asteroids would have produced large amounts of debris. As of the late 1960s there was not even an order-of-magnitude estimate for this region. Even a simple probe was considered valuable.[2]

Two of the eleven instruments onboard were designed to measure particles in the asteroid belt. The first consisted of thirteen panels with 234 pressurized cells mounted on the back of the dish antenna. When a cell was hit by a meteor, the skin would be penetrated and the gas would leak out. Pioneer 10 had thin-walled cells; Pioneer 11's cells had a thicker skin, which allowed detection of only larger particles. The second experiment used a meteoroid-asteroid detector, a package of four telescopes that would detect the light reflected from a particle. This could range from a dust-sized object passing a few feet away up to distant asteroids miles in diameter. When any three of the telescopes saw an object, the experiment logged an event. The object's relative size, distance, trajectory, and velocity could be calculated.

Pioneer 10 was launched on March 2, 1972, by an Atlas Centaur booster. When Pioneer 10 separated from the final stage, it was going 32,114 miles per hour, a velocity great enough to send it out of the solar system to wander endlessly among the stars. By June, shortly before it was to enter the asteroid belt, the cells had recorded 41 impacts. On July 15 Pioneer 10 crossed into the asteroid belt. By October, when it reached the midpoint of the belt, the impact rate had stayed much the same, with another 42 impacts recorded. This continued until mid-February 1973, when it cleared the asteroid belt. The meteoroid-asteroid detector showed that most of the particles were 0.1 to 1 millimeter in size, although a few were as large as 4 to 8 inches.

On April 5, 1973, soon after Pioneer 10 left the asteroid belt, Pioneer 11 was launched. By the time it left the asteroid belt, on March 20, 1974, a total of 20 impacts had been recorded, only 7 within the belt itself. Because of the thicker skin on its cells, these were from larger particles. These results indicated that the total number of particles was lower than had been estimated, and there were no swarms of particles that would endanger the spacecraft. The asteroid belt was not the barrier that had been feared.

Pioneer 10 made its closest approach to Jupiter on December 5, 1973. Pioneer 11 passed Jupiter on December 3, 1974, and was sent toward Saturn. It reached Saturn on September 1, 1979, and returned data and photos. Both probes continued out of the solar system.[3]

EARLY STUDIES

During the mid-1970s an asteroid sample return mission was discussed as a possibility for U.S.-Soviet space cooperation. Hannes Alfven of the Royal Swedish Academy of Science played the middleman between George M. Low of NASA and academician Roald Z. Sagdeev. The Soviets had a well-developed technology for sample returns and were developing plasma engines. The United States had the advantage in electrical power sources, which the plasma engines would require.[4]

The NASA analysis of the proposal noted that an asteroid sample return would require advances in propulsion, approach guidance, navigation, rendezvous, and docking (the low surface gravity of even a large asteroid would not allow a true landing). These were considered feasible, but engineering was not the major concern. The biggest problem was the lack of knowledge about the asteroids—there was no sound scientific basis for planning an asteroid sample return mission. If such a mission was to be more than a stunt, it must be preceded by simpler missions.[5]

Despite this, interest continued, and a conceptual breakthrough opened new possibilities. A spacecraft could use the gravity of one planet to fly on to the next target. It would not have been possible for the probe to carry enough fuel to make these maneuvers. Trajectory analyses from the Jet Propulsion Laboratory began calculating ever more complex missions. By early 1976 the lab had developed the Venus Earth gravity assist (VEGA). The probe would make flybys of Venus and Earth. Each time, it would gain speed. The possibilities for an asteroid mission were soon apparent. Direct ballistic trajectories to main-belt asteroids required launches with velocities between 4.9 and 8.4 kilometers per second. With VEGA, the required speed was reduced to only 3.16 to 6.32 kilometers per second. Almost any main-belt asteroid could now be reached.[6]

In the spring of 1980 NASA began looking at multiple asteroid rendezvous missions using a solar electric propulsion stage (SEPS). The mission would be launched in the late 1980s and cover six to eight asteroids over several years. The spacecraft would go into orbit around each asteroid for about 50 days, conducting mapping and mineralogical studies, before the SEPS would restart to move on to the next asteroid. There would be tremendous flexibility in the choice of asteroids to visit. Targets could even be selected after launch.[7]

The SEPS program was canceled, a victim of budget cuts. With funding for planetary missions now even tighter, attention shifted to spacecraft that could be developed for $150 million to $250 million, rather than the $1 billion cost of Galileo and Viking. Among the missions was a multiple asteroid rendezvous.[8] One way to keep costs down was to use existing technology, such as a TIROS weather satellite equipped with a scientific payload.[9] Some of the mission plans were quite involved. A plan for 1990 and 1992 launches called for flybys or rendezvous with six asteroids each, during a decade-long flight.[10]

At the same time, David Bender and Neal Hulkower of JPL's Mission Design Section were looking for asteroids easily accessible from Earth. Their initial target was 1943 Anteros, one of the near-Earth asteroids. Then, on February 27, 1982, as Eleanor F. Helin was photographing comet du Toit-Hartley, she discovered asteroid 1982 DB. The asteroid's long trail appeared near the comet as it flew past close to Earth. When Bender and Hulkower analyzed 1982 DB's orbit, they found it was the most accessible asteroid to a rendezvous mission.[11] It was later named 4660 Nereus, with a double meaning—Nereus was a Greek sea god, but the name is pronounced "Near Us."[12]

GALILEO

In the late 1970s work began on a Jupiter orbiter-entry probe mission named Galileo. In early 1984, with launch still more than two years off, JPL scientists began comparing Galileo's trajectory with the orbits of 3,600 main-belt asteroids. Depending on the exact launch date, Galileo could fly by either 1219 Britta or 1972 Yi Xing with only minor course changes. The first had been discovered by Max Wolf on February 6, 1932, and was among the last he found. The other was spotted by Purple Mountain Observatory on November 9, 1964, and was named for an ancient Chinese astronomer. Both were very small objects about which little was known.

Astronomers were urged to observe the asteroids to refine their orbits for navigation purposes and to measure their light curves, to determine the rotation rates and pole positions for camera targeting. Spectroscopic and polarimetry observations were also required to determine their surface composition, which would assist in analyzing the photo coverage.

Such observations were also seen as necessary to build support for the attempt.

The downside to a Galileo asteroid mission was that the added course corrections would cut into the fuel reserve at Jupiter, which could limit the number of observations. Another fear was based on the possibility that some asteroids had moons. There might be clouds of small debris orbiting the asteroids, which could destroy Galileo.[13]

The possibilities were further researched, and by the summer of 1984 a more impressive target became apparent, the asteroid 29 Amphitrite. Unlike the other two candidates, Amphitrite was a large object, nearly 200 kilometers in diameter. Its spectral signature indicated an S-type asteroid. The asteroid was thought to be either a heavily cratered remnant of the formation of the solar system or the core of a small planet that had melted, causing metal-rich minerals to sink to the center.

Galileo would be launched from a space shuttle on May 21, 1986, and make the Amphitrite flyby on December 6. It would then continue on to Jupiter, arriving on December 10, 1988. The detour to Amphitrite had a cost. The arrival at Jupiter would be delayed three months from the original plan, adding $10 million to $20 million to mission operations costs. The research plan at Jupiter would also have to be completely changed, and it was probable that some of the original research goals would not be met.

David Morrison, the chairman of NASA's Solar System Exploration Committee, sent a letter to NASA administrator James Biggs recommending the Amphitrite option. Morrison noted, "It will permit the U.S. to take the lead in asteroid studies rather than forcing us to come in second or third to the Russians or Europeans. . . . The alternative—to fly past this exciting object without even turning on the spacecraft instruments, thereby deferring our first exploration of the asteroids by nearly a decade—seems indefensible." A working group organized to study the option added, "Not only is the Galileo spacecraft well equipped to study Amphitrite, but . . . it would be difficult to design a better package of instruments for analyzing an asteroid during a flyby." On Christmas Eve 1984 Biggs gave his approval for the first asteroid mission.[14]

On January 28, 1986, less than four months before Galileo was to have been launched, the space shuttle *Challenger* exploded 72 seconds after liftoff. The shuttle program was grounded, and Galileo would not be launched for more than three years. The Centaur upper stage originally planned for Galileo was replaced by a less powerful solid-fuel rocket. It could no longer make a direct flight to Jupiter but would have to make the VEEGA maneuver (Venus Earth Earth gravity assist). Following an October 1989 launch, Galileo would fly by Venus in February 1990, Earth and the Moon in December 1990, asteroid 951 Gaspra in October 1991, and Earth and the Moon again in December 1992. A flyby of 243 Ida in August 1993 could then be made. Galileo would finally reach Jupiter on December 7, 1995.[15]

After more than a decade of effort, Galileo was launched from space shuttle *Atlantis* on October 19, 1989. During the final preparations, the launch team had to deal with hurricane Hugo, the San Francisco earthquake, two launch slips due to weather and a shuttle problem, and an attempt by three Florida antinuclear groups to get a court injunction to prevent the launch.[16] The first eighteen months of the mission went smoothly, but on April 11, 1991, Galileo's high-gain antenna failed to deploy fully, greatly reducing the speed at which photos could be transmitted to Earth. With the high-gain antenna, a single photo would take a few minutes to relay; the low-gain antenna, which had been used up to this point, required an hour and a half.[17]

951 GASPRA

Asteroid 951 Gaspra had been discovered by the Russian astronomer Grigory N. Neujmin on July 30, 1916, and was named for a Black Sea resort frequented by such literary figures as Tolstoy and Gorky. It was not until the Galileo mission, however, that Gaspra was subjected to intensive study. Ground-based observations would provide background data for analysis of the photos. In an echo of the early days of asteroid studies, some of the research was done by an amateur astronomer. In 1989 Claudine Madras, then a high school student in Newton, Massachusetts, wanted to do real scientific research. She contacted MIT asteroid specialist Richard Binzel and asked if she could work on a summer project. Binzel asked, "How old are you?" Madras replied, "Thirteen."

Despite her youth, Madras was assigned photometric observations of Gaspra, which were necessary to determine its rotation period. After learning how to obtain and analyze the data, she spent two cold nights observing with Lowell Observatory's 40-inch telescope (a major change from her department-store telescope at home). After several months of data reduction, she determined that Gaspra had a 7.04-hour period and was elongated and egg-shaped.[18]

Binzel and Noriyuki Namiki concluded that Gaspra was a fragment from a larger body and that its current surface should be much less than a billion years old. They expected a jagged body with a layer of dust only a meter or two in thickness (the latter because its escape velocity was only 10 meters per second; any impact would send surface debris off into space). From ground-based observations, Gaspra was known to be an S-type asteroid, a mixture of silicate material and metal. Near-infrared observations made by Jeffrey Goldader of the University of Hawaii indicated that metallic iron existed on Gaspra's surface, implying that Gaspra was originally part of the lower mantle of a larger body that had formed layers.[19]

Galileo flew by Gaspra on October 29, 1991. Because of the antenna problem, engineers were only able to confirm that the spacecraft had survived and that the scientific observations, using the full set of instruments, had been accomplished. Scientists believed that the playback of the photos would have to wait until the December 1992 flyby of Earth. The position of Galileo relative to Gaspra had been uncertain, so a pattern of pictures had to be taken to cover the predicted error.

Mission directors were confident about Galileo's navigation and decided to try an experiment. They believed they knew a frame had an image of Gaspra, but they had to guess where on the 800 × 800-pixel frame the image was and then where on the tape recorder that frame was. On November 6 a test transmission of 12 strips of pixels was made, which showed Gaspra dead-center on the frame. This was followed by transmission of the full 600-strip image of the asteroid. It took the 70-meter tracking antenna in Australia eight days to receive the four-color image.[20]

The image was released on November 14, 1991. Gaspra was 20 × 12 × 11 kilometers, a little larger than expected, but comparable in size to the Martian moon Demos and the nucleus of comet Halley (and the same size as the island of Oahu). As expected, it was an angular object, with a faceted shape, looking like a shark's head. The surface was covered by some three dozen craters; along with the smooth areas surrounding them, they indicated that the surface was geologically young. Project scientist Joseph Veverka estimated that Gaspra was the surviving fragment of a collision as recent as 300 million to 500 million years ago. The surface appeared to be covered by rubble, which gave it a soft look, a surprise given the earlier estimate. Enhanced images of the surface also showed a subtle pattern of fractures.

An exact mineralogical description would have to await the transmission from the near-infrared mapping spectrometer (NIMS). The initial data did confirm Gaspra was an S-type asteroid, made up of metal-rich rocks. The surface appeared uniform, which implied that it was an unaltered mix of rock and metal, without the layers of iron that had been suggested. Later, when Galileo's magnetometer data were transmitted, there was a surprise. Apparently the interplanetary magnetic field had shifted direction as the probe had flown by Gaspra, like the deflection of a compass needle due to a nearby iron deposit. The research team concluded that Gaspra was magnetized.[21]

243 IDA AND DACTYL

The Gaspra flyby completed, Galileo continued on its looping trajectory back toward Earth, then on to 243 Ida. This was a much larger asteroid—56 kilometers for Ida versus 20 kilometers for Gaspra—and was one of the approximately 150 members of the Koronis family. Binzel had extensively studied the Koronis family, concluding from their spins and shapes that the family could not have originated more than 1 billion years ago. Ida had been discovered on September 29, 1884, by Johann Palisa, and was named for one of the nymphs of Crete who nursed the young Zeus. From ground observations, it was known to be an S-type asteroid with a rotation period of only 4.63 hours and a very elongated shape.

The flyby of Ida was made on August 28, 1993. Galileo had been programmed to take 150 photos of Ida over a 5.7-hour period, a little more than a full rotation. However, Galileo was now too far from Earth for effective transmission of the photos. It was not until the end of the year that the data began to be transmitted and the first photo of Ida was released.

To overcome the antenna problem, JPL scientists used the same procedure they had with Gaspra. Only well-spaced narrow strips, called jailbars, were transmitted of each image. As Ann Harch was examining a jailbar image on February 17, 1994, she noticed something odd. One of the strips showed a bright object beyond Ida. The imaging team considered and eliminated the possibility of a planet or a star. On February 23, another team was examining data from the NIMS and noticed something funny. There was only one possibility.[22]

As the new image of Ida formed line by line on the screen at JPL, the long debate over the existence of asteroid moons came to an end. To one side of Ida was a small spherical object. Word quickly spread about the new moon. The March 7, 1994, issue of *Aviation Week and Space Technology* carried a brief item saying that "Galileo seems to have spotted the first moon of an asteroid ever sighted."[23]

The formal announcement was made on March 23, 1994. Michael J. S. Belton of Kitt Peak Observatory noted that the discovery "probably means they are quite common." Making only two asteroid flybys and having one of them turn up a moon implied there were more. It was given the provisional designation 1993 (243) 1—"1993" for the year of observation, "243" after Ida's number, and "1" as the first moon discovered. Its nickname was Baby Ida. In September the IAU approved the permanent name Dactyl, derived from beings in Greek mythology called Dactyli, who lived on Mount Ida with Zeus. Dactyl itself was 1.2 × 1.4 × 1.6 kilometers in size and roughly spherical in shape.

Dactyl had implications for the formation and history of the asteroids. The first result was an estimate of Ida's mass and composition—once the size and orbital period of Dactyl were known, the mass of Ida itself could be calculated. When this was divided by Ida's volume, its density could be found. The density, along with the NIMS data, would indicate Ida's composition and internal structure. The 150 photos, taken over more than one Ida day, allowed its shape (and therefore volume) to be reconstructed.

Determining Dactyl's orbit, however, proved to be more difficult. Galileo had made its flyby in nearly the plane of Ida's equator, while Dactyl's orbit was inclined only 8.5 degrees. Although the moon appeared in 47 photos, the resulting perspective was so poor that a wide range of orbits could fit the imaging data. Assuming that Dactyl was barely held by Ida's gravity, a very eccentric orbit of 80 × 8,000 kilometers, with a period of one year, was possible with Galileo happening to catch Dactyl close to Ida. At the other extreme, a nearly circular orbit of 82 × 95 kilometers would also fit the data. This orbit had a period of 27 hours. Computer simulations by Jean-Marc Petit, Richard Greenberg, and Paul Geissler indicated that if the low point of Dactyl's orbit was within 70 kilometers of Ida's center, the moon would become unstable and would either crash into Ida's surface or be flung out into space. Petit noted that an exception was possible if Dactyl's orbit was an even multiple of Ida's spin rate. In such a case, stable orbits could probably exist with low points under 70 kilometers. (The 27-hour orbit was just short of a 6-to-1 ratio.)

With all the uncertainties, Ida's density was calculated to be between 2.2 and 3.0 grams per cubic centimeter. That eliminated the possibility that Ida was composed of stony-iron material, which required a density of 5 grams per cubic centimeter or more. Even if Ida was a loose collection of iron-rich fragments, with a third of its volume empty space, its density would be too high to match the Dactyl results.

The Galileo data highlighted a debate that had been under way since the early 1970s. Ida and Gaspra were both S-type asteroids. This type showed a slight reddish color and moderate reflectivity and was the most common type in the inner asteroid belt. Ida, however, was more distant than was typical for S-types, which spectroscopically resemble both ordinary chondrites and stony-iron meteorites. Ordinary chondrites were the most common meteorites found on Earth, so it was expected that they would also be found in abundance among the asteroids.

Subsequent research, however, indicated that S-type asteroids were not a particularly good match for ordinary chondrites. They are too red, and their infrared absorptions due to olivine and pyroxene are too weak. Ida's reddish color resembles that of an object with considerable amounts of metallic iron, but this was ruled out by the Dactyl data. Clark R. Chapman of the Planetary Science Institute had long suggested that S-type asteroids underwent space weathering, which changed their appearance, and the photos of Ida support his idea. Several areas, including around fresh craters, were less red and showed stronger spectral absorption features than the rest of Ida. These areas were not an exact match with ordinary chondrites, but they were closer than the rest of Ida.

Galileo's magnetometer, as with the Gaspra flyby, detected a shift in the magnetic field around Ida. The situation was complex, but it appeared that Ida's interior had high electrical conductivity. This did not mean that there were large deposits of nickel-iron, however.[24]

Faced with proof of an asteroid satellite, astronomers now had to figure out how Dactyl could have been formed and then survived. Ida was a splinter of an original body about 125 miles across that was shattered in a collision. The fragments became the Koronis asteroid family. How could one body have ended up in orbit around another following such an event? It was speculated that when the original body shattered, some of the fragments traveled along identical paths, flying side by side. As time passed, their gravitational attraction caused one to go into orbit around the other. Computer modeling indicated that if a 110-kilometer body shattered, there was a 50 percent chance that at least one of the larger fragments would attract a moon.[25]

Such a scenario, in which Ida and Dactyl are not mother and daughter but big sister and little sister, is supported by their differing surface compositions. Although they are similar in color, Ida is olivine with only a small amount of orthopyroxene, while Dactyl is roughly equal parts olivine, orthopyroxene, and clinopyroxene. If they had come from different parts of the original body, such subtle differences would be expected. Binzel noted that the members of the Koronis family also show subtle color differences, similar to those of Ida and Dactyl. These could be due to slight differences in composition, the space weathering effect, or both.

The surfaces of both Ida and Dactyl showed evidence of violence. Ida is covered with hundreds of craters, dozens of boulders 30 to 150 meters across, and numerous grooves. Dactyl shows more than a dozen craters, the largest 300 meters across (about a fourth of its diameter). Both surfaces seem equally battered, and the amount of cratering would indicate an age for both surfaces of at least 1 billion years.[26]

This contradicted extensive dynamic studies of the Koronis family by Binzel and others that indicated a younger age. Binzel suggested the heavy cratering could be a result of the formation of an asteroid family. Possibly Ida had been hit by smaller fragments as the original body broke up. After the family formed, the individual members would also have had fairly close orbits. Fragments knocked off one asteroid could then hit another. It was only later that the orbits became well separated. Both processes could have resulted in an excessive number of craters. Binzel warned, "Just because you see lots of craters, you shouldn't automatically assume it's old."

The final question was why Dactyl still existed—calculations indicated that an object the size of Dactyl should not last longer than 100 million years before being destroyed by an impact. Possibly it had been shattered, but the fragments, still in orbit around Ida, had been gravitationally attracted back together. Its rounded shape, unlike that expected for a collision fragment, could indicate it was a loose collection of rocks held together by gravity.[27]

The years following Galileo's discovery of Dactyl fueled the belief that asteroid moons were common. Petr Pravec at Ondrejov Observatory, in the Czech Republic, observed the small near-Earth asteroids 1994 AW1 and 1991 VH. The first, about one kilometer in diameter, showed a dimming that indicated a moon about half its size in a 22.4-hour orbit. The asteroid 1991 VH showed an eclipse every 33 hours from a moon about 40 percent of the primary object's size. The third discovery was made by astronomers at the European Southern Observatory. While observing 3671 Dionysus, they found a 28-hour cycle determined to be from a moon. Dionysus was also small, with a diameter of one kilometer.

The existence of these asteroid moons was inferred from the light curves, but the first direct image of an asteroid moon from a ground-based telescope was announced in October 1999. Astronomers from the Southwest Research Institute, using the Canadian-French-Hawaii telescope on Mauna Kea, Hawaii, spotted a moon orbiting the main-belt asteroid 45 Eugenia. Eugenia's diameter is about 215 kilometers; the moon, designated S/1998(45)1, is estimated to be about 13 kilometers in diameter. It is in a circular orbit about 1,190 kilometers away from Eugenia, with a period of five days. The data indicated that Eugenia has a density only about 20 percent greater than water. William J. Merline, the team leader, said, "A picture is emerging that some asteroids are real lightweights." Clark Chapman, a team member, added, "Either these objects are highly porous rubble-piles of rock, or they are mostly water ice."

The Southwest Research Institute discovery was the first from a search program of 200 asteroids. Merline commented, "If more satellites are found, it will revolutionize our understanding of the makeup of asteroids." The overall number of asteroids with moons is suggested by the craters on Earth. Of 28 large impacts, 3 are double craters, roughly 1 in 9. A similar ratio of single to double craters is found on the Moon, Venus, and Mars.[28]

CLEMENTINE: NEW DIRECTIONS IN SPACE RESEARCH

By the time Galileo successfully reached Jupiter on December 7, 1995, the project had been under way for two decades, at a cost of $1.4 billion.

Such high costs and long development times could not be supported politically. Clearly a new direction was needed. It would come from an unusual source. In 1983, as Galileo awaited launch, President Ronald Reagan proposed the Strategic Defense Initiative (SDI). Intended to build a defense against Soviet ICBMs, it would involve both space-based and ground-based interceptors, and they would require high-precision detection and tracking systems. Such systems would have to be tested to determine the effects of radiation and the heat and cold of space.

Since the end of the Apollo program, NASA had wanted to conduct a detailed mapping and mineralogical survey of the Moon. The final three Apollo flights had made a partial survey, but the areas covered were limited. An unmanned spacecraft put into a polar orbit around the Moon could cover the entire surface, but repeated attempts over nearly two decades failed to win approval for such a flight.

In 1990 it was realized that SDI's test goals and NASA's scientific interests coincided. In 1991 the Strategic Defense Initiative Organization (SDIO), which managed the various SDI programs, began discussions with NASA about flying the system tests on a Moon and near-Earth asteroid mission. In January 1992 SDIO formally notified NASA that it would be undertaking the mission, later named Clementine. NASA organized a thirteen-member science team, headed by Eugene M. Shoemaker, in April 1993.

The Naval Research Laboratory (NRL) was selected to build the Clementine spacecraft, working in conjunction with the Goddard Space Flight Center and JPL. NASA's Deep Space Network provided tracking for Clementine. The NRL built Clementine in only two years, at a cost of only $55 million (not including the sensor costs, which were provided by SDIO). The package of sensors, built by the Lawrence Livermore National Laboratory, consisted of a CCD camera for use in visible and ultraviolet wavelengths, near- and far-infrared cameras, and a combined high-resolution CCD camera and laser-ranging system (LIDAR). They would conduct a complete survey of the Moon at eleven different wavelengths. Ten were selected by the NASA science team to detect the mineral composition of the Moon and the near-Earth asteroids 1620 Geographos and 3551 1983 RD.[29]

Clementine was launched on January 25, 1994, by a refurbished Titan IIG ICBM, from Vandenberg Air Force Base. This was the first planetary mission ever launched from Vandenberg. Clementine made two loops around Earth, then went into lunar orbit on February 21. It began the lunar survey a week after entering orbit and continued until April 23. By the time it was finished, scientists had more than 1.5 million images of the Moon, providing complete coverage of the lunar surface.[30] Clementine data also seemed to show pools of water ice in several permanently shadowed craters at the Moon's south pole.[31]

With the lunar segment of Clementine's mission now complete, its main engine was fired on May 4 to send it on toward Geographos. Clementine was scheduled to arrive in August 1994 for the first close look at an Apollo asteroid. Geographos was believed to be highly elongated, possibly dumbbell-shaped, 3 to 4 kilometers long and 1.5 kilometers wide. Scientists thought that Geographos was made up of several large fragments weakly bonded together by gravity.[32] Assuming Clementine had enough fuel, it would continue on to the asteroid 3551 1983 RD, with the flyby planned for October 1995. This was a much smaller body, with a diameter of only a kilometer.[33]

On May 7, three days after leaving lunar orbit, disaster befell Clementine. During a rehearsal for the Geographos flyby, four of its attitude control thrusters fired for 11 minutes, depleting the steering gas supply and leaving it spinning at 80 revolutions per minute. By mid-May the engineers realized the Geographos flyby was out of the question. Like the Western heroine it was named for, Clementine was now "lost and gone forever."[34]

The Galileo and Clementine missions were very different, but together they established asteroids as priority targets for space science missions. As planetary missions underwent a resurgence in the late 1990s, an asteroid mission was one of the first flown.

NEAR

In 1992, as Clementine was being built, NASA administrator Daniel Goldin announced the start of the Discovery program. These planetary missions would cost only $150 million each, have well-defined objectives, use as much existing technology as possible, and take only three years from approval to launch. One of the first missions approved was the

Near Earth Asteroid Rendezvous (NEAR) mission, which would orbit an asteroid for a year of observations.[35]

NEAR was designed by the Applied Physics Laboratory at Johns Hopkins University. Approval was given in December 1993, with the launch scheduled for February 1996. This 27-month period was shorter than any planetary mission in the previous two decades. NEAR carried a multispectral CCD camera, a near-infrared imaging spectrograph, an X-ray–gamma ray spectrometer, and a magnetometer. A laser altimeter would be used for the delicate maneuvering into orbit around the asteroid, and a radio science experiment would be used to determine the asteroid's gravitational field. As with Clementine, some of the technology was originally developed for military use.[36]

The target for NEAR was the asteroid 433 Eros. Eros was an Amor asteroid, which means it crosses the orbit of Mars but does not quite reach the orbit of Earth. Light-curve data indicated that Eros was cylindrical in shape and had a rotational period of 5.27 hours. The asteroid also rotated on its side. Spectroscopic observations showed a typical S-type asteroid, except that the two sides of Eros seemed to have differing mineralogical content, possibly indicating that a recent impact had spalled a large chunk off one side, exposing fresh surface material. Radar data also seemed to indicate a recent impact.[37]

As with Galileo, there would be a flyby of an asteroid along the way. Originally, this was to be 2968 Iliya in August 1996. During the refining of the trajectory, however, the mission designers discovered that if NEAR was launched before February 27, 1996, it would pass within 1,200 kilometers of 253 Mathilde. At this range, Mathilde would be 2.9 degrees across, compared to 0.5 degrees for the Moon. The flyby could be done with no expenditure of fuel or significant added cost. Approval came swiftly.

The contrast between the two objects could not be more striking. Eros was an S-type asteroid (as were Gaspra and Ida), but Mathilde was a C-type, with a dark surface made up of carbon-rich material unchanged since the formation of the solar system. Mathilde was also large, at about 60 kilometers, compared with Eros, estimated to be 40.5 × 14.5 × 14 kilometers. Eros was oblong in shape; Mathilde was believed to be more spherical.[38] One of the strangest features of Mathilde was its slow rotation rate. Eros rotated once every 5.27 hours. Mathilde, in contrast, rotated once every 17.4 *days*. How an asteroid could rotate so slowly was unexplained.[39]

Mathilde was a main-belt asteroid discovered by Johann Palisa on November 12, 1885. Unlike the mythological name given to Eros, the origin of "Mathilde" was conventional. It was named for the wife of astronomer Moritz Loewy by V. A. Lebeuf, who calculated the asteroid's orbit. Lebeuf was a staff member at the Paris Observatory while Loewy was the vice director.

243 MATHILDE

NEAR's ambitious development schedule was completed on time, and in early February 1996 it was ready for launch. The first attempt was made on February 16, but problems with the range safety equipment forced a halt with an hour remaining in the launch window.[40] The second try proved more successful, and at 3:43 p.m. on February 17, the Delta II booster lifted off and sent NEAR on its long journey. NEAR would travel three years and a total of 1.3 billion miles before it reached Eros.[41]

Seventeen months after launch, NEAR began final preparations for the Mathilde flyby. It took place on June 27, 1997, and lasted for 25 minutes. In that time NEAR took 534 images using the multispectral CCD. NEAR was almost twice as far from the Sun as Earth, and because of the reduced amount of light hitting the solar panels, the camera was the only one of the scientific instruments turned on.[42]

The first image, received just before 10:00 a.m., showed a section of the potato-shaped asteroid with a 19-kilometer-deep impact gouge, a hole greater than the volume of Mount Everest. Joseph Veverka, the imaging science team leader, observed, "The degree to which the asteroid has been battered by collisions is astonishing." Donald K. Yeomans of JPL said, "Mathilde is an asteroid with a very tortured past." Eugene Shoemaker, on seeing the photo, said, "It's more crater than asteroid." At least five craters larger than 20 kilometers in diameter were on the 60 percent of the surface imaged. Raised crater rims indicated the debris had quickly fallen back to the surface after the impact. Straight sections on several crater rims indicated the presence of large faults or fractures in the asteroid material. The large number of craters suggested that Mathilde's slow rotation was due to these huge impacts.[43]

The surface reflected only 3 percent of the sunlight shining on it, which made it twice as black as charcoal. Veverka said, "We knew that C-asteroids are black, but we did not expect their surfaces to be as uniformly black and colorless as Mathilde's surface turned out to be. This global blandness is an important clue telling us that asteroids such as Mathilde are made of the same dark, black rock throughout because none of the craters, which are punched deep into the asteroid, show evidence of any other kind of rock."

At the same time, it was not clear how Mathilde could have survived such impacts without shattering. Each of the objects that struck it was as large as several kilometers. The depths of the resulting craters were a significant fraction of Mathilde's own diameter. The gravitational effect of the close flyby on NEAR's own trajectory gave a possible explanation. Yeomans said, "Preliminary results suggest that Mathilde is much less dense than we had thought."[44] Shoemaker said, "If it is just a loose collection, you could get very large craters." This would be true even if the objects were small. Mathilde measured about 59 × 47 kilometers, somewhat smaller than expected. It was also more rounded than Gaspra or Ida. Mathilde's gravity was so low that an object that weighed one ton on Earth would weigh only four pounds on its surface. NEAR searched for any possible moons, but none were found.[45]

TOWARD EROS

NEAR spent almost another year coasting toward Eros. Finally, on December 20, 1998, NEAR's rocket engine ignited for the 20-minute burn that was the initial step to place it into orbit around Eros. One second into the firing, however, a computer software problem shut the engine down. NEAR then went into a violent spin, causing a loss of communications with tracking stations. For the next 27 hours, communications remained blacked out, and the mission seemed lost. Finally, the computer was able to halt the spin and restore communications. In the process 30 kilograms of maneuvering fuel had been burned. There was no chance to make the planned orbital maneuver, but a flyby could still be made, with a second orbital attempt later. Ann Harch, Maureen Bell, and Scott Murchie began writing the software for the flyby. It was written and debugged in 24 hours, rather than the six months normally required, then transmitted to NEAR with 8 minutes to spare.[46]

On December 23 NEAR flew past Eros, taking 1,026 images from as close as 3,830 kilometers. They showed an elongated object with a large impact crater, looking as if a huge bite had been taken out of the asteroid. There was evidence of color differences in the surface, which implied a diverse composition. There were fewer craters on the surface than on Ida, indicating a younger surface. An elongated ridge stretched across some 20 kilometers of its surface. The images also showed that Eros was 33 × 13 × 13 kilometers, slightly smaller than the radar studies had indicated. No moon was seen in the images.

As NEAR flew by Eros, the asteroid's gravitational attraction caused a slight change in its trajectory. From this, a density of about 2.7 grams per cubic centimeter was calculated, a value within the range derived for Ida but twice that of Mathilde. Joseph Veverka noted that both the ridge and the higher density "suggest that Eros is a homogeneous body rather than a collection of rubble" like Mathilde.

Finally, after almost a four-year voyage, at 10:33 a.m. EST on February 14, 2000, Valentine's Day, NEAR fired its engine and successfully went into orbit around Eros, the asteroid named for the Greek god of love. The initial images after the engine burn showed a heavily cratered surface, which also featured grooves, layers, and house-sized boulders scattered across its surface. Andrew Cheng, of the Applied Physics Laboratory at Johns Hopkins University and NEAR lead scientist, said, "There are tantalizing hints that the asteroid has a layered structure, like a sheet of plywood. These layers appear to be very flat and appear to run end to end. This could come about if Eros was once part of a larger body, perhaps a fragment of a planet." The large number of craters implied that the surface was about 2 billion years old. The only smooth area was the Saddle, a wedge-shaped groove in its surface. It might be either a new feature or an old surface that had been covered by a slide. Initial measurements indicated the surface was composed of pyroxene and olivine, which are commonly found in meteors. With NEAR in orbit, the density measurements made during the flyby were refined. They indicated that Eros has a density of 2.4 grams per cubic centimeter, about that of

Earth's crust. The data confirmed the initial estimate that Eros is a solid body, rather than a rubble pile like Mathilde.[47]

DS 1

The Deep Space 1 (DS 1) mission is part of JPL's New Millennium program to test new low-cost spacecraft technologies for use on twenty-first-century flights. DS 1 carries a solar electric ion propulsion system, similar to the ion propulsion system originally proposed in the late 1970s. DS 1 also has an onboard autonomous navigation system, called AutoNav, which will make decisions about spacecraft trajectories and targeting with little input from Earth-based controllers. The capabilities of the AutoNav system and the ion propulsion system have been likened to a car driving itself from Los Angeles, California, to Washington, D.C., then parking itself in a preselected parking space, and using only one tank of gas in the process.

DS 1 was originally planned for a launch in July 1998 on a Delta II booster. In January 1999, DS 1 was to fly by the asteroid 3352 McAuliffe (an Amor asteroid named for Christa McAuliffe, killed in the space shuttle *Challenger* explosion). Then in April 2000 DS 1 was to fly by Mars, to be followed only two months later with an encounter of comet West-Kohoutek-Ikemura.[48] However, delays in delivery of the spacecraft's power system and an ambitious flight software schedule did not leave sufficient time to complete the checkout for a July launch. The mission was rescheduled for an October launch, which meant that a new set of targets would be needed.[49]

After looking at 100 flyby possibilities, the near-Earth asteroid 1992 KD was selected. The flyby was scheduled for July 28, 1999, with DS 1 passing as close as 33,000 feet from its surface. Should the mission be extended, it could then go on to the comet-turned-asteroid 4015 Wilson-Harrington in January 2001, then to comet Borrelly in October 2001.[50]

DS 1 was successfully launched on October 24, 1998. After some initial problems with the ion engine, it ran smoothly for more than 850 hours between early November 1998 and early January 1999. The AutoNav system was tested in its fully autonomous mode, controlling the ion engine and the spacecraft's course. The other new technologies aboard DS 1 were also tested, and observations were made of stars and Mars to test the scientific equipment.[51]

As the July 28 flyby date neared, a name was being selected for 1992 KD. Eleanor Helin and the Planetary Society had selected "inventors" as the theme for the name, reflecting DS 1's advanced technology. More than 500 entries were submitted from around the world, with the winning entry by Kerry Babcock, a software engineer at Kennedy Space Center. Babcock proposed that 1992 KD be named for Louis Braille, inventor of the raised dot system that enabled blind individuals to read. Babcock's citation read, "Louis Braille invented the Braille language so those who could not see could obtain knowledge and explore through the written word. Likewise, asteroid Braille provides knowledge about our universe and its origin to the people of Earth, who through Deep Space 1, are also able to explore and discover what previously they could not 'see'." The asteroid was formally renamed 9969 Braille the day before the flyby.[52]

As with NEAR, there was a problem during final preparations. Only sixteen hours before the flyby, there was an electronic shutdown aboard DS 1. The final commands were radioed up to the spacecraft only 4 minutes before contact would have been lost. When the data were radioed back to Earth following the flyby, it was apparent that another problem had occurred. They showed only the blackness of space, rather than images from as close as 26 kilometers away. Only four black and white images were returned: two taken about 70 minutes before the closest approach and two lower resolution photos taken about 15 minutes after the pass. They showed a peanut-shaped object, about 2.2 × 1 kilometer in size, possibly made of two or more chunks. There was little in the way of surface detail. Flight controllers speculated that Braille was dimmer than expected, or that the camera was less sensitive. As a result, the camera was unable to lock onto the asteroid.

Although the images were a disappointment, the spectral data made up for it. Braille was a close match for the basalt-covered surface of Vesta. Both are V-type asteroids. The spectral similarities imply that Braille is a fragment knocked off Vesta, possibly when the large impact crater was formed. While Vesta remained in the main belt, Braille was sent into an orbit that caused it to be influenced by Jupiter's gravity, which then sent the asteroid into a new orbit passing close to Earth. The flyby completed the DS 1 mission. On July 30 the ion engine began a three-month-long burn, to direct DS 1 toward its two additional targets.[53]

FUTURE ASTEROID MISSIONS

If current plans hold, the next two asteroid missions will be to 4660 Nereus, the Apollo asteroid identified two decades before as the easiest to reach from Earth. They would not be U.S. or Russian missions, however, but rather Japanese, and the first commercial deep space mission. The Japanese Muses-C mission would be launched in January 2002 by an M-V booster from the Kagoshima Space Center in Japan. The Muses-C spacecraft would land on Nereus in September 2003.[54] A NASA-built nanorover, weighing only 2.2 pounds, would then move across its surface, looking at rocks, observing their grain structure, examining them with a spectrometer, and collecting samples. Although it has four wheels, its normal mode of travel would be to hop from place to place. Michael Newell of JPL said, "Our wheels can spin us right off the surface." If the nanorover lands upside down, the wheels' strut assembly can right it. The nanorover's samples would be loaded into a return capsule, and the spacecraft would blast off to begin the return to Earth. The capsule would parachute to a landing in January 2006.[55]

In August 1998 SpaceDev Inc. announced it had also selected Nereus as the target for its NEAR asteroid mission. The project is a commercial operation—rides for scientific instruments are to be sold to governments and companies, and the data gathered would then be sold by SpaceDev. The revised plan proposed a launch on April 3, 2001, with the spacecraft remaining in the Earth-Moon system until January 12, 2002. During this period, Earth and Moon studies, such as looking for ice deposits on the Moon, could be made. The spacecraft would then leave Earth's gravity and encounter Nereus on about May 12, 2002. The primary mission would be completed by mid-June 2002.[56]

The farthest frontiers of the solar system would be explored by the next proposed future mission, the Pluto-Kuiper Express. The mission had its start with a postage stamp. On October 1, 1991, JPL engineers and stamp collectors gathered at the Von Karman Auditorium to see the unveiling of ten stamps commemorating U.S. planetary exploration. They showed the nine planets of the solar system and the Moon, along with the unmanned spacecraft that had explored them—all of them, that is, except Pluto. The stamp said, "Not Yet Explored."

JPL employee Robert Staehle took this as a challenge and talked with trajectory engineer Stacy Weinstein about a high-speed, low-cost Pluto mission.[57] The Pluto Express, as it was first called, was originally to be a direct-trajectory mission to Pluto, which would have required the Russian Proton, making the launch costs too high. Multiplanetary gravity assists came to the rescue. The spacecraft would use a Venus Venus Venus Jupiter gravity assist (VVVJGA) trajectory to reach Pluto: it would swing by Venus three times, each time gaining a little more speed, which would send it farther out into space, and then encounter Jupiter, which would send it the rest of the way to Pluto.[58]

As the project was beginning, the first of the Kuiper Belt objects was discovered, and it was realized that the planet Pluto was only the largest of these icy minor planets. The mission was expanded and renamed the Pluto-Kuiper Express. The spacecraft could be launched in 2003 and fly by Pluto and its moon Charon in 2013. This accomplished, it could then continue on to one or more Kuiper Belt objects.[59]

The asteroid space missions of the 1990s and beyond changed the human perception of asteroids. No longer were they only dim points of light in a telescope eyepiece or long trails on a photographic plate. We had, at last, seen them close up, as worlds in their own right.

7

THE NAME'S THE THING!

In fact, it is safe to say that the situation has degenerated to the point of absurdity. Asteroids have been named after girlfriends, financial supporters, cats, and computers. For a traditionalist like myself, it seems a pity that the naming of asteroids has become so trivialized.

Charles T. Kowal, 1988

A little nonsense now and then is relished by the wisest of men.

Anonymous

THE NAMES GIVEN TO asteroids have been the subject of both interest and controversy. The process has been beset from the start by politics, personal clashes, and differing views about what is proper. In the nearly two centuries since the first asteroid name was argued over, they have honored myths of the past and the future, heroes and villains, astronomers and scientists, places, plants, and other things. The meanings of some names have been lost to history. It is by naming these orbiting mountains that we make them part of the human experience.

CERES FERDINANDEA AND VICTORIA

When Piazzi announced the discovery of the first asteroid, there was intense interest in what it would be named. Napoleon discussed the question with Pierre Simon de Laplace on the battlefield. Laplace preferred the name Juno (later used for the third asteroid). Other French astronomers wanted to name it Piazzi, after its discoverer. German astronomers were split between Juno or Hera (the Greek equivalent of Juno, later used for asteroid 103).[1]

Piazzi himself was firm—he named the object Ceres Ferdinandea, "Ceres" after the patron goddess of Sicily and "Ferdinandea" after Ferdinand IV, king of both Sicily and Naples, who had taken refuge in Palermo from Napoleon's invading army. Naming the first asteroid after a reigning monarch was not popular, particularly in France and countries allied with Napoleon. Piazzi would not bend. He said later, "I have the full right to name it in the most convenient way to me, like something I own. I will always use the name Ceres Ferdinandea, nor by giving it another name will I suffer to be reproached for ingratitude towards Sicily and its king."[2] Despite his insistence, the name was soon shortened to Ceres.

The three asteroids that followed were given solid classical names: Pallas, Juno, and Vesta. When the long drought in asteroid discoveries ended in 1845, it made sense to continue using the names of female Greek or Roman mythological figures. Thus the fifth asteroid was named Astraea, after the goddess of justice. There were no problems until the discovery of the twelfth asteroid on September 13, 1850, by the English astronomer J. R. Hind. He selected the name Victoria, after the Roman goddess of victory. It also happened to be the name of the queen of England at the time.[3]

The criticism was led by B. A. Gould, editor of *Astronomical Journal.* He wrote, "Such nomenclature is at variance with established usage, and is liable to the objections which very properly led astronomers to reject the name Ceres Ferdinandea." For several years after, Gould used the name Clio for the asteroid, which had been suggested as an alternative name by Hind.

Support for the name Victoria was led by William C. Bond of Harvard College Observatory. He stated, "Victoria was the daughter of Pallas, and one of the attendants of Jupiter, and, therefore, the name appears to fulfill the required conditions of a mythological nomenclature." For his part, Hind wrote in a letter to *Astronomical Journal:* "The name Victoria was submitted to the approbation of astronomers on mythological grounds, and not exclusively as marking the country where the discovery was made. I foresaw the objections which you [Gould] have advanced. . . . I would at once reject any name that is not founded in mythology. It seems to have been forgotten that Her Majesty's name is derived from the goddess, who cannot thereby lose her celestial rights."

In the end, "Victoria" received the support of most astronomers. Partly in response to the controversy, European astronomers discussed the naming issue and decided to continue the practice of using female mythological names.[4] Events would show, however, that the question of names was far from resolved.

THE ATTACK OF THE CLASSICISTS

Further trouble came with the twentieth asteroid, discovered on September 19, 1852, by A. de Gasparis at Naples and the following night by Jean Chacornac at Marseilles. It was called 20 Massalia by Benjamin Valz, using the Greek name of Marseilles, and that was a problem. It was the first asteroid name not taken from classical mythology. (This was also the first asteroid to be given a number; it was Valz who originated the practice of adding the asteroid's number before the name.) The second such break with mythological practice was 21 Lutetia, discovered on November 15, 1852, by Hermann Goldschmidt. The name was selected by F. J. D. Arago. As Goldschmidt lived in Paris and Arago was the director of the Paris Observatory, it seemed logical to give the asteroid the Latin name for Paris. A third asteroid with a nonclassical name was 25 Phocaea. Like 20 Massalia, it was discovered by Chacornac and shared ties with his home in Marseilles. Phocaea was a maritime town of Ionia in Asia Minor. About 600 B.C., a group of its inhabitants fled and established what became Marseilles.

Two more of Goldschmidt's discoveries, made in 1858 and 1859, would also stray from the classical road. The asteroid 45 Eugenia was named for the French empress Eugenia de Montijo, the wife of Napoleon III and the first human to be so honored. It was followed by 54 Alexandra, named for Baron Alexander von Humboldt, who explored South America, Mexico, and Siberia. Although Goldschmidt made it clear he was naming it after Humboldt, the nonclassical name was accepted, as Alexandra was also the daughter of the king of Troy.

In 1860 and 1861 the use of classical versus nonclassical names became a major issue in the astronomical community. The fifty-ninth asteroid discovered was named Elpis, after the Greek word for hope. Urbain Leverrier, the director of the Paris Observatory, did not want to use names and proposed a new procedure: the number would be followed by the discoverer's name. The proposal ran into immediate opposition from a number of noted astronomers. At the same time, two asteroids discovered by E. W. Tempel on March 4 and 8, 1861, added more fuel to the debate. Valz named 64 Angelina in honor of an astronomical station established near Marseilles by Baron Franz Xaver von Zach. For the sixty-fifth asteroid, Tempel gave the right to name it to Bavarian minister K. A. von Steinheil, who selected Maximiliana, after the king of Bavaria, Maximilian II, which raised another storm of protest.

The most outspoken of the classicists was Karl T. R. Luther, founder of Bilk Observatory near Dusseldorf. He argued, "As long as people believe it appropriate to give special names to celestial bodies like stars, comets, the moons of Saturn and Uranus and even for the mountains of the Moon, it seems also appropriate to adhere to names from classical mythology, since a mere number could easily lead to mistakes and misunderstandings. Unclassical names, however . . . are not tenable in the long run, this being the alternative, one should rather turn to the numbers alone." He concluded by suggesting that if discoverers lacked the self-discipline to use proper names, then the individuals who calculated asteroid orbits, called computers at the time, should take action. "Classical names are necessary, unclassical names are rejected; the computers have the right to [replace] unclassical by classical names."

Luther's dogmatic attitude did not pass uncontested. Steinheil responded by observing sarcastically, "What advantages should be . . . in choosing only classical names? Are the new planets to remind us of having gone to a classical secondary school?" He continued by noting a difficulty: "It seems that Dr. Luther has not taken into account the practical side of his demand. Who should have the right to [replace] unclassical names if two or even more orbital computers are working on each planet—each of them?" Goldschmidt sided both with the use of names and the classicists, but in a much less dogmatic way than Luther. He wrote, "I vote . . . against the repeal of names, which would cause a great confusion. Some displeasing names are not worth the trouble of annoying their patrons. A planet is all number—the name is the only poetic part of it, and this shall now become again a number without there being a real cause to do so."

The result was a split decision. Leverrier dropped his attempt to introduce a new system in January 1862. Asteroid 59, which had been the trigger for his attempt, received not one but two names, Elpis in Germany and Olympia in France and England. The duel names remained in use for a short time, before 59 Elpis was finally accepted. Although 64 Angelina stood, Maximiliana was rejected and replaced by 65 Cybele, after a Phrygian goddess.

For a few years peace reigned within the astronomical community, as classical names were used exclusively. By the late 1870s, however, several incidents of backsliding raised the classicists' ire. Asteroid 125 Liberatrix was named to honor the end of the Prussian occupation of France, 141 Lumen for a book by the French astronomer Camille Flammarion, and 154 Bertha and 169 Zelia for his sister and a niece.[5]

This was too much for Luther, and in 1878 he again demanded use of classical names: "The names recently becoming a more than colorful mixture . . . it seems very advisable to return to the old use of preferring classical, mythological names. Allusions of any kind should be avoided—for the sake of the honor of science. . . . A stricter adherence to classical names will hopefully help to shield the growing number of minor planets from increasing indifference."[6]

Luther's comment about allusions was an apparent reference to several cases of mythological names with double meanings. Asteroid 55 Pandora, named for the woman who released all the evils in the world, also referred to a bitter dispute between B. A. Gould and the Dudley Observatory trustees. Asteroid 102 Miriam was named for the sister of Moses. C. H. F. Peters selected the name so he could tell a theological professor, who he thought was too pious, that Miriam was also a mythological figure. Despite the classicists' renewed attacks, asteroids continued to be given improper names. In defiance of the edict against current events, 185 Eunike was named in honor of the signing of the treaty ending the Russo-Turkish War (the name translates as "Happy Victory"). Asteroid 494 Virtus (the personification of virtue) was named at the request of Flammarion, who observed that if virtue disappeared from Earth, it would be nice to find it in the heavens.

What was probably the low point in the first century of asteroid names was represented by 250 Bettina, the first and only asteroid to be sold. The Austrian astronomer Johann Palisa was having trouble finding names for his asteroids, and he needed money for an eclipse expedition. To solve both problems he placed an ad: "Herr Palisa, being desirous to raise funds for his intended expedition to observe the total solar eclipse of Aug. 29, 1886, will sell the right of naming the minor planet No. 244 for 50 pounds." There was no immediate response, and in the meantime he discovered two more asteroids, 248 and 250. Palisa named each one in turn but left the last open for a patron. Finally Baron Albert von Rothschild, of the Austrian banking family, bought 250, naming it after Baroness Bettina von Rothschild.

Although selling an asteroid was extreme, others were named for astronomical benefactors. Asteroid 323 Brucia was named for Miss Catherine Wolfe Bruce, who paid for Max Wolf's telescope at the Heidelberg Observatory. It was the first numbered asteroid discovered photographically, and the Bruce telescope has been used to discover more numbered minor planets than any other in the world.[7]

Again, the classicists raised objections. George Chambers, in his 1913 book *Story of the Solar System,* said, "the most fantastic and ridiculous names have in many cases been selected, names which in too many instances have served no other purpose than that of displaying the national or personal vanity of the astronomers who applied them to the several planets. The French are the great offenders in this matter."[8]

By the end of the nineteenth century the Astronomisches Rechen-Institut (ARI) in Berlin had responsibility for asteroid names. In 1899 the ARI gave its official position: "There is reason to ask the discoverers not to offend against the rule of choosing female names. . . . Male names will not be accepted." With several asteroids named for cities and sponsors, the policy could not be maintained. The result was a linguistic compromise—the names of cities, male astronomers, sponsors, and others were given the Latin feminine suffix "ia" or "a." The asteroid named for the French village of Herment, for example, was 346 Hermentaria. The sole exception was the use of male mythological names for Earth-crossing and Trojan asteroids.

The practice of adding a feminine ending to male names had largely faded out of use by World War II. The Minor Planet Center, which took over responsibility for numbering and naming asteroids in 1947, made the change official: "The custom of attaching feminine endings to masculine names has had numerous exceptions in the past. Names which are submitted will not be rejected or modified if they are masculine."

There were also several informal naming practices in the early part of the twentieth century. Asteroids with large orbital eccentricities and inclinations were given names with four letters. One example was 1508 Kemi, named for a Finnish city. Another practice was the selection of names based on the provisional designation. Examples of this were 574 Reginhild (1905 RD), 580 Selene (1905 SE), and 593 Titania (1906 TT). Both practices later largely died out.

DOCUMENTING ASTEROID NAMES

At the twentieth IAU General Assembly in 1988, Edward Bowell and several others suggested the establishment of a "Study Group on the Origin of Minor Planet Names." It would list the names of all numbered asteroids and document their meanings. (During the early years, there was no requirement to explain the source or meaning of the names.) The result, published three years later, was the *Dictionary of Minor Planet Names*. At that time, there were 5,012 numbered asteroids. When the third edition was published in 1997, the total had grown to 7,041. By 1999 the total of numbered asteroids had reached 10,000.

The study group found that the names fit into nineteen categories. Despite their initial impression, the names did not fall into an "astronomers' cemetery" or a "female sky." In the 1991 edition men's names totaled 1,717 versus 509 women's names, and 132 asteroid names had unknown meanings (most of which were women's names that could not be identified as belonging to a specific person).

Not surprisingly, astronomers' names made up the largest single category, with 888, many of which involved asteroid studies. As the number of asteroids neared 1,000, the opportunity was taken to commemorate the individuals who had set it all in motion. Johann Bode was honored with 998 Bodea; 999 Zachia was named for Baron Franz Zach; 1000 Piazzia was in honor of Giuseppe Piazzi, discoverer of 1 Ceres; 1001 Gaussia recognized Carl Gauss, whose mathematical efforts led to the recovery of 1 Ceres a year later; and 1002 Olberia was named for Heinrich Olbers, the discoverer of 2 Pallas and 4 Vesta. This tradition has been continued: 2000 Herschel honors William Herschel, who undertook some of the earliest asteroid studies, and 4000 Hipparchus is named for an early Greek astronomer.

Other asteroid researchers remembered in this manner include 1578 Kirkwood, after Daniel Kirkwood, who first described the structure of the main asteroid belt. Asteroid 1999 Hirayama is named for Kiyotsugu Hirayama and his work on asteroid families, and 1006 Lagrangea is for

Joseph L. Lagrange, whose solution of the three-body problem predated the discovery of the Trojan asteroids. Asteroids named for discoverers include 1111 Reinmuthia, for Karl Reinmuth, who discovered the most asteroids, and 2005 Hencke, for Karl Hencke, who broke the long drought when he spotted 5 Astraea and 6 Hebe. The asteroid 1614 Goldschmidt is named for Hermann Goldschmidt, the artist-turned-astronomer who discovered 14 of the first 70 asteroids, and 914 Palisana is named for Johann Palisa, the last of the visual observers. The asteroids 827 Wolfiana and 1217 Maximiliana are both named for Max Wolf, who pioneered the use of photography for asteroid discoveries. An unusual name is 1152 Pawona, which is for both Johann Palisa and Max Wolf, in recognition of their joint work. This asteroid, originally designated 1925 SF, was the last discovered by Palisa before his death. It was not seen again until January 1930, when it was recovered and recognized.

Ironically, 1303 Luthera was named for Karl T. R. Luther, who led the classicists' fight for mythological names in the late nineteenth century. He might not have been pleased by the honor.

Mythological terms made up the second largest group, with 458 (just over half the astronomers' total). Although mythological names predominated in the nineteenth century, by the end of the twentieth they were the exception. The Amor asteroids include 4055 Magellan (after both the explorer and the spacecraft); the Hilda asteroids include 334 Chicago, 1578 Kirkwood, and 1268 Libya. Only the Trojan asteroids remain mythologically pure. With the large number being discovered, there have been fears that names will run out. It was half-jokingly suggested that astronomers would be reduced to using names like First Trojan Soldier and Second Trojan Soldier.

Close behind, with 393 asteroids, are cities, harbors, and buildings. The city names often reflect some involvement with astronomy. In the late nineteenth and early twentieth century it was common practice to name an asteroid after a city that played host to a scientific meeting. For instance, 449 Hamburga was named during the 1901 festival of the Mathematical Society of Hamburg, Germany, and 723 Hammonia was also named for Hamburg, during the twenty-fourth annual meeting of the Astronomische Gesellschaft in 1913. In later years there was a shift in names. Because of the work of Finnish astronomers during the 1940s and 1950s, a great many asteroids were named for Finnish cities and ports. Still others were named after African cities, because of the discoveries made by South African astronomers, and Chinese cities appear thanks to the work at Purple Mountain Observatory.

Several of these names have been obscured by time. Both 764 Gedania and 1419 Danzig were named for the Free City of Danzig ("Gedania" was based on the city's Latin name). After World War II, when the city was renamed Gdansk, the asteroids retained the earlier name. In August 1991, following the failed hard-line coup, Leningrad reverted to its pre-Communist name, St. Petersburg. Asteroid 2046 Leningrad, however, retained its Soviet-era name.

The next group, with 306 members in the IAU compilation, recognizes scientists who are not astronomers. It includes Nobel laureates such as 1332 Marconia, for 1909 physics prizewinner Marchese Marconi, one of the pioneers of radio; 1069 Planckia, after Max Planck, the 1918 physics winner who did early work in quantum physics; 2001 Einstein, for 1921 winner Albert Einstein, who revolutionized physics with the theory of relativity; and 3581 Alvarez, which jointly honors Luis W. Alvarez, the 1968 physics laureate, and his son, Walter Alvarez. As a team they discovered a thin layer of iridium in worldwide rock deposits, leading to the theory that an asteroid or comet impact was responsible for the extinction of the dinosaurs. As with astronomers, the millennial asteroids have honored scientists. The asteroid 7000 Curie is named for Marie Curie, the chemist and physicist, and 8000 Isaac Newton honors the man who developed the laws of gravity and invented calculus.

Another 225 asteroids are named for relatives and friends. Asteroid 153 Hilda was named for the eldest daughter of astronomer Theodor von Oppolzer, 228 Agathe for his youngest daughter, and 237 Coelestina after his wife. Asteroid 631 Philippina was named for Philip Kessler, as a gift on the occasion of his engagement. It is even possible to trace the family trees of astronomers in asteroid names. The Shoemaker Dynasty of asteroids includes sixteen relatives of Carolyn and Eugene Shoemaker.

Close behind, with 212 entries, are asteroids named for countries, provinces, and islands. Among the countries are 232 Russia, 241 Germania (Germany), 293 Brasilia (Brazil), 329 Svea (Sweden), 1071 Brita (Great Britain), and 1112 Polonia (the Latin name for Poland). Asteroids named for U.S. states include 341 California, 439 Ohio, 793 Arizona, 1602 Indiana, and 3124 Kansas. There is also 1193 Africa, as well as 1278 Kenya and 1279 Uganda. Befitting its size, population, and role in asteroid discovery, China has three asteroids named for it, 139 Juewa (Star of

China's Fortune), 1125 China, and 3789 Zhongguo (a transliteration of the Chinese). Among the island asteroids are three named for Hveen, the island on which Tycho Brahe built his observatory: 379 Huenna (the Latin name), 499 Venusia, and 1678 Hveen. Given the 200-year history of asteroid naming, it is not surprising that time has taken its toll on some of this group. When 371 Bohemia was discovered in 1893, the namesake region was part of Austria. After World War I Bohemia became part of Czechoslovakia (honored by 2315 Czechoslovakia). Following the end of the Cold War, Czechoslovakia peaceably split into two new states. Not so fortunate was the namesake of 1554 Yugoslavia—that country disintegrated in a brutal civil war. The breakup of the USSR also had its impact in the sky. Originally, 1284 Latvia, 1541 Estonia, 2577 Litva (Lithuania), 1709 Ukraina, and several other asteroids honored parts of the Soviet Union; after August 1991 they became independent states. Finally, 1197 Rhodesia honors a country now called Zimbabwe.

The next two groups of asteroid names are related: writers, with 156 entries, and literary figures, which trail behind with 136. Not surprisingly, great writers of the European Renaissance appear in the sky, including 2984 Chaucer, 2985 Shakespeare, and 2999 Dante. More contemporary writers honored include 2810 Lev Tolstoj, 2625 Jack London, 2675 Tolkien (author of *Lord of the Rings*), 3412 Kafka, and 3508 Pasternak, after Nobel prizewinner Boris Pasternak. Popular writers are also included: 1931 Capek honors Czech playwright Karel Capek, best known today as author of the play *RUR* (Rossum's Universal Robots), a term that is now a staple of science fiction; 2730 Barks is named for comic book writer and illustrator Carl Barks, whose stories often featured space exploration and minor planets (more than twenty years before, he had touched on the idea of "rubble pile asteroids"). Occasionally, there is a more whimsical reason for the honor. Asteroid 2817 Perec is named for French writer Georges Perec, who wrote a 300-page novel titled *La Disparition,* which did not use the letter "E." "This eccentricity," the naming citation noted, "would seem to suit him to studies of minor planets."

Literary figures immortalized by asteroids include 643 Scheherezade, named for the fictional wife of an oriental king and the narrator of the tales of the Arabian Nights. A more contemporary fiction character is honored by asteroid 1640 Nemo, after the captain of the submarine *Nautilus* in Jules Verne's science fiction novel *Twenty Thousand Leagues under the Sea.* Sometimes the names have a double meaning. Asteroid 3552 Don Quixote was named for the Miguel de Cervantes character who attempts to bring back the age of chivalry. The asteroid's extended orbit shows a similar long-term erratic nature.[9]

One of the most touching stories to contribute an asteroid name is that of Joshua Cole, hero of the 1923 novel *Cole of Spyglass Mountain.* The naming not only reflects the story itself but also what the book meant to a father and his son. In the novel Joshua Cole runs away from his brutal father, is caught, and sent to a boy's reformatory called the House of Refuge. Cole loses his name, becoming "5,635." During his years there, a staff member introduces him to astronomy and to Lowell's ideas of a Martian civilization. When Cole leaves, he is given a telescope. He makes his way west, setting up the telescope by the roadside to sell looks at the stars to passersby—until the telescope is stolen.

Cole finally reaches California, finds work on a railroad tunnel, and saves enough money to buy a new telescope. Nearby is a hill with the clearest, darkest skies that he has ever seen. He sets up a homestead on the site, which he names Spyglass Mountain, and begins observing Mars, trying to confirm Lowell's theory. Cole's evil father, however, has learned that Cole is due a very large inheritance. To prevent him from getting it, the father has a neighbor try to burn down Cole's house and take shots at his observatory.

All the time, Mars is drawing closer to Earth. Finally, on the night of June 18, as Cole is observing Mars, for a brief moment the canals become clear and unmistakably artificial. Then the neighbor opens fire again, and Cole is hit. He survives, but at first no one can confirm his report. Finally, an observatory reports photographing the features. The story ends with the line, "First to report discovery, Cole of Spyglass Mountain famous in a night."

In 1960 David H. Levy became interested in astronomy. One night at dinner, his father told him about reading *Cole of Spyglass Mountain* when he was fourteen years old and described the story. He was impressed by the story of a young man and his dream, and in the years to follow, he would ask his son if he had ever found a copy of the book. In that time David Levy became a successful amateur astronomer, comet hunter, and author. As his father grew older, he showed signs of Alzheimer's disease. It worsened until he no longer seemed to remember that David was his son. A few months before he died, he was walking with David when he turned and asked, "Did you ever find *Cole of Spyglass Mountain*? What a story that was!"

It was not until May 1997 that David Levy was finally able to find a copy of the book. He also checked a list of numbered asteroids and noted that a particular one had not yet been named. He wrote the discoverer, Bobby Bus, suggesting an appropriate name. Bus agreed, and in December 1997 the name 5635 Cole was announced.[10]

Asteroids named for amateur astronomers total 119 in the IAU compilation. In some cases amateur astronomers have achieved distinction in fields far removed from the night sky. Asteroid 3125 Hay honors William Hay, an accomplished planetary observer. Hay discovered a white spot on Saturn in 1933. During the 1930s and early 1940s, however, he was better known to the public as English music hall comedian and film star Will Hay. Amateurs historically have had a leading role is the discovery of comets, and their accomplishments are honored in the sky. Asteroid 3850 Peltier was named for Leslie C. Peltier, one of the most distinguished U.S. amateur astronomers of the twentieth century. Between 1925 and 1954 he discovered 12 comets. He also discovered several novae and over 62 years made 132,123 variable star observations. His 1965 autobiography *Starlight Nights* introduced a new generation of amateurs to the night sky.

Every asteroid name tells a story, some of great accomplishment, others of sadness—2134 Dennispalm tells both. C. Dennis Palm was an active amateur astronomer and night assistant at Palomar Observatory. Born in 1945, he was only 29 and a half years old (to the day) when he died. Yet in that short time, he had accomplished everything he had wanted to do, except build his own small observatory. As a teenager, he would sit in an ice cream store, look up at Palomar Mountain, and say to himself, "Someday I'll work up there." Two years after his death, on Christmas Eve 1976, Charles T. Kowal discovered 1976 YB, which he named for Palm.[11]

The next category has proven the most difficult and controversial—historical and political figures, with 117 asteroids. The problems that first appeared with Ceres Ferdinandea have continued to the present day. The great ideological struggle that lasted from the start of World War I to the fall of the Soviet Union in 1991 was documented in the sky. Asteroid 852 Wladilena is named for Vladimir Ilyich Ulyanov, known to history as Lenin. (The asteroid's name used the first syllables of his name and was given a few months after Lenin's death in 1924.) Other asteroids were later named for Communist officials: 1277 Dolores was for Dolores

Gomez Ibarruri (La Pasionaria), a founder of the Spanish Communist Party, and 1550 Tito was named for Yugoslavian president and World War II resistance leader Josip Tito.

Paul Herget, who directed the Minor Planet Center from 1947 until 1978, established a liberal attitude toward political names. He said, "In the past, names have been rejected on the grounds of political connotations. This policy will not be continued in the future."[12] Still, the issue resurfaced with asteroid 2807 Karl Marx. Frederick Pilcher, chairman of the physics department at Illinois College, suggested in late 1983 that the name be rescinded. Pilcher wrote, "This extension of the Cold War into the sky is intolerable."

Brian Marsden, who took over direction of the Minor Planet Center when Herget retired, responded, "I deplore as much as [Pilcher] does the use of the minor planet names for purposes of propaganda, but occasional appearances of political or ultra-nationalistic names are far from new, and many such names have most decidedly been proposed by the discoverers of the planets concerned. . . . There are also instances where discoverers have named minor planets to honor political figures or ideas, only to want to change the names when the figure or ideas became out of favor! Such changes have not been permitted, and I think that most astronomers would not wish the names of minor planets to be as impermanent as those of airports, streets, and other manmade constructions."

Marsden explained that the flood of new asteroids eligible for naming demanded a new procedure. IAU Commission 20, which has responsibility for the Minor Planet Center, felt that the rules developed by the IAU Working Group on Planetary System Nomenclature would be too restrictive. (The body approved names for craters and other features on planets and moons.) They explicitly banned the use not only of the names of political or religious figures but also of living persons.

To deal with the problem, in 1982 the commission established the Minor Planets Naming Committee to examine each proposal for suitability. It was initially made up of the president and vice president of IAU Commission 20 and the director of the Minor Planet Center. Marsden explained that several months had passed before all three members were convinced that 2807 Karl Marx was an appropriate name. The citation that accompanied the naming described Marx's accomplishments in what the committee thought was a factual manner and was seen as better than the original Soviet proposal.

The basic problem, as Marsden noted, was that "one could probably find someone somewhere and at some time in history who would object to the name of any minor planet, from 1 Ceres on." Pilcher accepted Marsden's arguments, responding, "Less hurried contemplation shows that this [rescinding of political names] is a bad precedent, and will lead to chaos if carried on extensively. In a future and saner age the historical records would have to be searched and the original names reassigned."[13]

In 1985 the IAU General Assembly passed a resolution attempting to clarify the rules. It stated, "Names glorifying individuals or events principally known for their political or military activities or implications are considered unsuitable unless at least one hundred years have elapsed since the individuals died or the events took place."

Far less controversial is the category of mountains, rivers, and seas. A few examples of the 108 asteroids include 1317 Silvretta and 1584 Fuji, 1149 Volga and 1345 Potomac, and 4042 Okhotsk. Of the nineteen categories, this is the last to contain more than 100 entries in the 1991 IAU compilation. The remaining six all total less than 100 each. First among these runners-up are the 92 asteroids named for institutions and observatories, including such schools as 694 Ekard (Drake spelled backward) and 2906 Caltech. Observatories recorded in the sky include 2322 Kitt Peak, 2343 Siding Springs, 2460 Mitlincoln (after MIT's Lincoln Lab, an early center for radar astronomy), 990 Yerkes, and 991 McDonalda (McDonald Observatory with the "a" ending).

Following close behind, with 88 asteroids, is the category of composers and musicians, comprising an unofficial poll of astronomers' tastes in music. The names fall into two categories. First are classical composers, including such asteroids as 1814 Bach, 1815 Beethoven, 3975 Verdi, 1034 Mozartia, 3590 Holst (composer of "The Planets"), 3910 Liszt, 3992 Wagner, 4559 Strauss, and 4532 Copland.[14] The other category was inaugurated by the Fab Four asteroids, 4147 Lennon, 4148 McCartney, 4149 Harrison, and 4150 Starr. These asteroids were discovered during 1983 and 1984 by Brian A. Skiff and Edward Bowell of Lowell Observatory.[15] Later 4305 Clapton was named for rock guitarist Eric Clapton, best known for the songs "Layla" and "Wonderful Tonight." While rock and roll was making inroads in the sky, the only country and western asteroid was 4779 Whitley, named for singer Keith Whitley, who died tragically at the peak of his career.[16]

It is a matter of taste, and as such, even this seemingly noncontroversial category attracted objections. Earlier objections were based on the honor of science, appeals to tradition, and legitimate questions about political suitability, but in 1990 the critics turned to name-calling. A staff writer with the *Washington Times* newspaper became offended when the four asteroids were named for the Beatles. He referred to the two astronomers as "Mad scientists . . . whose heads have been in the clouds too long . . . whose judgment is as bad as their musical taste." The names remained.[17]

Trailing far behind, with only 58 asteroids, are those named for plants and animals, including 1095 Tulipa and 1320 Impala. This group is largely the result of Karl Reinmuth's efforts. He published a long list of 66 new asteroid names between 1009 Sirene and 1200 Imperatrix, including 28 consecutive names of plants and flowers starting with 1054 Forsytia, named after a member of the olive family with yellow bell-shaped flowers. In 1902 Max Wolf discovered two asteroids that were named 482 Petrina and 483 Seppina, after his dogs, Peter and Sepp. Astronomer Z. Vavrova also named 2474 Ruby after his dog.

The IAU study found that naming asteroids after plants and animals had fallen out of fashion; the trend in the 1980s was to use acronyms. Among these 41 asteroids are 3654 AAS, honoring the American Astronomical Society; 3568 ASCII, named for the American Standard Code for Information Interchange, the standardized computer characters; and 2848 ASP, for the Astronomical Society of the Pacific on the occasion of its centenary. Although 1725 CrAO looks like a chemical formula, it actually stands for Crimean Astrophysical Observatory. The name 3325 TARDIS stands for Time and Relative Dimensions in Space, a device with the outward appearance of a British police telephone box that was prominent in the long-running science fiction TV series *Dr. Who.* As the number of asteroids approached 5,000, special names were selected. Asteroid 4999 MPC was named for *Minor Planet Circulars,* which announced new discoveries and the names approved for asteroids; 5000 IAU was for the International Astronomical Union itself; and 5001 EMP stood for *Ehfemeridy Malykh Planet* (Ephemerides of Minor Planets), issued by the Institute of Theoretical Astronomy in Leningrad (now St. Petersburg).

Painters and sculptors account for 32 asteroid names, among them 2919 Dali, 3000 Leonardo, 3001 Michelangelo, 4221 Picasso, and 4457 van Gogh.

The final category, called curiosities, is the smallest, with only sixteen entries in the 1991 list, but the most entertaining. The IAU study noted, "Somewhat to our regret, there are not many entries in the category of curiosities. This is a wide field for imaginative discoverers in future."[18] Brian Marsden seconded this, saying later, "I just like to see imaginative names. Most of the names submitted are terribly boring." Asteroid 2500 Alascattalo is named for a mythical Alaskan beast produced when a moose and a walrus were crossed. A parade, lasting four minutes and extending a block down an alley, is held every year at three minutes past noon on Alascattalo Day, the first Sunday after the third Saturday in November.[19]

2309 MR. SPOCK

The intractable nature of the asteroid-naming controversy can be seen in the story of one example. In early 1985 astronomer James Gibson and his wife, Ursula, were tired of seeing asteroids named for political figures, friends, or children—people who had contributed nothing to astronomy but had been immortalized in the sky. As they wrote later, "Why, our cat had done more for astronomy than many such people."

Many people shared this view; they knew Brian Marsden did, and they suspected Tom Gehrels felt the same way. Marsden had met their cat and understood how its companionship had added to life at isolated observatories in Argentina and South Africa. Marsden suggested they write up a citation for one of the ten asteroids James Gibson had discovered between 1971 and 1985. It would not only honor their cat but would also emphasize how the asteroid-naming situation was getting out of hand. They hoped that the controversy it would spark might start a wide-ranging discussion of what was an appropriate name. The citation was submitted and, to their surprise, was approved. It read: "2309 Mr. Spock: Named for the ginger short-haired tabby cat (1967–) who selected the discoverer and his soon-to-be wife at a cat show in California and accompanied them to Connecticut, South Africa and Argentina. At El Leoncito he provided endless hours of amusement, brought home his trophies, alive or dead, and was a figure of interest to everyone who knew him. He was named after the character in the television program *Star Trek* who was also imperturbable, logical, intelligent and had pointed ears."[20]

The response was soon in coming—some astronomers loved the

whimsical name, while others were furious. At the 1985 IAU General Assembly, Tom Gehrels made a blistering attack on the idea of naming asteroids after household pets. IAU Commission 20, which oversees asteroid names, subsequently passed a resolution discouraging such pet names. There was, however, no outright ban.

James and Ursula Gibson wrote later that had they been in attendance, they would have agreed with many of Gehrels's objections. They felt it would have been better, however, if the resolution had not merely discouraged pet names but also reinforced the tradition of naming asteroids after those who had made contributions to astronomy, however significant or unusual. "That would still leave us free," they wrote, "to name asteroids for composers and musicians whose works make the long hours in the dome pass more pleasantly, and for night assistants, laboratory helpers, computer experts, and other vital support persons." To their disappointment, there was no broad discussion of asteroid names.[21]

In the end the names given to asteroids are a reflection of the human experience. One should not expect uniformity of opinion about asteroid names any more than one would expect it about books or movies or music. It would be unfair to ban specific types of names that have been permitted in the past. If the discoverer's wishes are to be vetoed, it should take more than the word of an aggrieved editorial writer. Throughout the two centuries of debate over correct names, there were very few who believed that the naming of asteroids should, itself, stop. It would be unfortunate if the wonder were lost and in the future we were reduced to a sterile list of numbers and letters.

With all due respect to those involved, however, in this case something was overlooked. Asteroids historically have been named for mythological figures. Mr. Spock *is* a mythological figure.

8

3043 SAN DIEGO: THE UNWANTED HONOR

America's Finest City

San Diego city motto

READING THROUGH A LIST of asteroid names, one finds the entry 3043 San Diego. Its brief citation gives no indication of the story of how the asteroid got its name, or of the hostile act against astronomy it now represents. The reader's initial reaction to these events will be one of disbelief, that such people cannot exist, that the political and civic leaders in one of California's largest cities could not really do and say such things. But they did.

TWO CITIES: TWO DIFFERENT PATHS

This story began in the early years of the twentieth century, as astronomy attempted to understand the structure of the universe. The quest brought George Ellery Hale to the peak of Mount Wilson. Here he built a 100-inch telescope, then the largest in the world. But Hale had greater ambitions. He saw the Mount Wilson Observatory as the means to transform Pasadena, then a small town of some 15,000 people at the base of the San Gabriel Mountains, into a center of scientific research. Hale joined local groups, and they became enthused with his dream. The changes Hale brought to Pasadena were profound; the California Institute of Technology (Caltech), the Huntington Library, and the Pasadena civic center made the city an intellectual and cultural center.

But Hale was still looking ahead. In the mid-1920s he began laying the groundwork for a 200-inch telescope.[1] It would not be built atop Mount Wilson, however. The lights of rapidly growing Los Angeles were already beginning to affect the 100-inch telescope. Hale looked to the remote northern back country of San Diego County, to Palomar Mountain. The area was selected with the expectation that the city of San Diego, some 70 miles away, would never pose a threat.[2]

San Diego was a very different city from the Pasadena that Hale, Mount Wilson, and Caltech had made. While Pasadena was being transformed into a scientific, intellectual, and cultural powerhouse, San Diego was attempting to decide what kind of city it would be. The question boiled down to the slogan "smokestacks versus geraniums," that is, an industrialized city versus a resort enclave. There was no San Diego counterpart of Hale, to give direction or a unifying vision of the city's future. As a result, the issue of smokestacks versus geraniums was never settled. But if San Diego could not decide what it was, it *could* decide what it was not. San Diego would not be Los Angeles. San Diego became a smug cul-de-sac, a sleepy little Navy town with little connection with the outside world.[3]

These two different cities became linked at 3:00 a.m. on September 21, 1931, when Caltech, San Diego County, and local ranchers signed an agreement to build the 200-inch telescope atop Palomar Mountain. It took place by candlelight, while a storm raged outside the old cabin. Local boosters were proud of the event and even wanted to change the name of the site to San Diego Mountain. Delayed by World War II, the 200-inch Hale telescope was finally dedicated on June 3, 1948.[4]

OPENING SHOTS

In response to the increased energy costs of the 1970s, California mandated that San Diego's mercury-vapor streetlights were to be replaced by more energy-efficient ones. The two possibilities were low-pressure sodium (LPS) lights, which are yellow in color, and high-pressure sodium (HPS) lights, which emit a pink color. As a test, limited numbers of both types of lights were installed in different areas of San Diego. Eight areas were equipped with LPS lights.

One of the test areas was the adults-only neighborhood of Seven Oaks, in the Rancho Bernardo area of San Diego. The LPS lights in Seven Oaks sparked an extreme reaction. A 63-year-old resident complained that he could not sit on his patio anymore. "I felt like the lights were intruding." Two elderly sisters complained that the LPS lights were penetrating their home, illuminating the living room and a back bedroom. "They're terrible, just terrible," one of the sisters said. They and others complained bitterly about the lights to Bob Kyle and the Rancho Bernardo Town Council, saying they were intolerable. Kyle and the council decided that LPS lights were unacceptable not only in Rancho Bernardo but anywhere in San Diego, and they went to the area's representative on the San Diego City Council, Bill Mitchell.

Mitchell later said, "I had one person come in to me and get down on her knees and pray to God that those yellow lights would go away." One man complained that he had taken his wife out to dinner "and she looked like a cadaver" under the LPS lights. Another person told Mitchell, "If you have caps on your teeth, they come out looking black, like you have no teeth." What had begun as the complaints of a very few people became a crusade by Mitchell to protect the city from what he called "bug lights."[5]

The LPS lights glowed yellow because they emitted virtually all their light in a single narrow band. It was therefore a simple matter to filter out the LPS lights, and telescopes could operate with few problems. The light from HPS streetlights, however, was spread across the whole visible spectrum, making them impossible to filter out. Light pollution from the growth of San Diego in the previous two decades had already reduced the 200-inch telescope to the equivalent of a 140-inch telescope. If the 27,000 San Diego streetlights were converted to HPS lights and the surrounding areas followed San Diego's lead, then the area's growth rate would mean Palomar Observatory would be blinded within ten years. Also affected would be San Diego State University's observatory on Mount Laguna and the San Diego Astronomy Association's observatory in eastern San Diego County.[6]

STRUGGLE AND DEFEAT, 1981–1983

At first, Mitchell and Rancho Bernardo insisted that LPS lights would increase crime and accidents. The yellow light from LPS lamps changed colors. Reds, for example, appeared brown. This, they argued, would hamper police work. In contrast, HPS lights were described as a deterrent to crime. LPS supporters were quick to note that San Diego had the lowest ratio of police to population of any major California city, and that HPS also changed colors. The LPS lights illuminated a wider area with a more uniform level of light.[7] LPS lights had already been installed in several California cities, and neither crime nor accident rates had changed.

Mitchell also pressed the theme of economics. The HPS lights had a life expectancy of 24,000 hours, and LPS bulbs were expected to last 18,000 hours. This, Mitchell claimed, meant the HPS bulbs had lower maintenance costs and were more economical. Mitchell also said that the city manager had recommended against LPS lights on economic grounds; in fact, the city manager's office had found a cost savings that favored the LPS lights. Their energy costs were $100,000 lower than HPS, which more than countered an estimated $20,000 in higher LPS maintenance costs.[8]

Mitchell did gather some allies. During the late 1970s and early 1980s San Diego spent considerable money trying to revitalize its decaying downtown area. The results were mixed. Downtown remained populated by street people—a drunk sleeping off a binge under a park bench, empty bottle at his side, while others wandered aimlessly.[9] The Center City Development Corporation and downtown businesses blamed LPS lights for the area's shortcomings and warned that people would be afraid to come downtown unless HPS lights were installed. They wanted an exemption for the downtown area allowing the HPS lights.[10]

The streetlight issue wended its way through hearings before the Public Services and Safety Committee and the full San Diego City Council. The first vote, on November 23, 1982, ended in a 7 to 1 defeat for Mitchell and Rancho Bernardo. A second vote, on March 21, 1983, reaffirmed the decision to install LPS lights. The vote this time was 4 to 3, with council members Dick Murphy and William Jones joining Mitchell. All that remained was for the city council to approve the LPS contracts.

In the spring of 1983, however, there was a massive turnover on the

San Diego City Council. Mayor Pete Wilson had been elected to the U.S. Senate, and three council seats had changed hands. The new mayor was Roger Hedgecock, a young, liberal Republican whose slow-growth policies ran counter to the local conservative establishment. Hedgecock was also an avid surfer and had a surfboard made with the city seal on it.[11]

The city council was to meet a final time on the lighting issue in June 1983 to approve the awarding of $1.2 million in contracts for the first phase of the LPS conversion program. Mitchell asked for a rehearing because Mayor Hedgecock and Councilwoman Gloria McColl had not participated in the earlier decision. Both McColl and Hedgecock had already expressed their opposition to the LPS lights.[12]

With this behind-the-scenes move, Mitchell surprised the LPS supporters and assured his victory before the meeting was ever held. The council voted on June 21, 1983. Many of the speakers on both sides had already appeared before the council and its subcommittees half a dozen times. Michael Anderson, president of the San Diego Astronomy Association and a supporter of LPS lights, told a reporter that he had already lost more than $1,000 in wages so far "because this council has not been able to say no to Mr. Mitchell and make it stick."[13] The meeting itself was only a formality. When the vote was taken, it was the expected 5 to 4 in favor of the HPS lights—Mitchell, Hedgecock, McColl, Murphy, and Jones versus Bill Cleator, Ed Struiksma, Mike Gotch, and Uvaldo Martinez.

After the vote Mitchell was exuberant, saying, "Truth will out," and adding, "I started out as the lone vote against the LPS system, so it is very satisfying to now have a majority of the City Council taking my position." Bob Kyle of Rancho Bernardo said, "I'm tired but I'm happy. I feel like I just won a tough race."[14]

FIGHTING FOR A RIGHTEOUS CAUSE

In the wake of Mitchell's victory it was felt by some that Caltech had taken a hands-off attitude in its dealings with San Diego. It was San Diego State University's astronomy department that had the day-to-day dealings with the city council.[15] Following the vote, however, Caltech took a direct role in reversing the decision.

To build support, Caltech invited journalists to see the 200-inch telescope in operation. Warren Froelich, a writer with the *San Diego Union* newspaper, spent a cloudy night talking with Matthew A. Malkan, a 27-year-old astronomer studying quasars. Malkan bluntly told Froelich what the city council's vote for HPS lights meant: "It would have the same effect as taking a sledgehammer and knocking out half of it. Virtually everyone here is using this telescope for faint object searches. What the city of San Diego proposes means disaster. It's the beginning of the end." Caltech also hired San Diego attorney Paul Peterson, who had experience with city council rules and regulations, to work on the issue on a day-to-day basis.[16]

The efforts to reverse the council's decision were not limited to Palomar or astronomers. David Lasser was one of many individuals throughout San Diego and the world who became involved on their own. Lasser was retired and lived in the Seven Oaks area of Rancho Bernardo, the same area that had set the events in motion. In August 1983 Lasser appeared before the Rancho Bernardo Community Council. He told them, "There should be an impartial commission, appointed by the city council, so that all the facts could be out in the open." Bob Kyle curtly brushed Lasser off: "I don't want to see this whole issue voted on again by the city council. I don't want the apple cart completely upset. That would be chaos. I prefer the idea of an orderly progression."[17]

This was not the first time David Lasser had taken an unpopular stand. In 1931 he founded the American Interplanetary Society, the first such organization in the United States, and he served as its first president. At the time, the United States was sinking deeper into the Great Depression. Lasser formed a local union in 1933, which went national the following year as the Workers Alliance of America. He was denounced on the floor of the U.S. House of Representatives by Martin Dies (D-Tex), who said, "This fellow Lasser is not only a radical, he is a crackpot. He is affected with mental delusions and thinks we can travel to the Moon." Congressmen laughed at such an impossibility.[18]

LPS supporters, whether organized or acting independently, emphasized the economic price that San Diego would pay. Councilman Mike Gotch, the strongest supporter of LPS lights, told *Science* magazine, "San Diego is being watched by a number of high-tech companies. They are waiting to see what kind of support we give to the academic and technical communities before they move here."[19] A letter to the editor of the *San Diego Union* made the same point in the bluntest possible terms:

"When I attend out-of-state conferences, I constantly hear comments about how small-minded and provincial the government and people of San Diego must be to allow the virtual destruction of this awesome observatory. . . . I have even heard high-placed executives comment that maybe San Diego is too hostile to science to consider locating their high-tech industries there."[20]

The most devastating commentary on San Diego's action came from Carl Sagan. He wrote an article for the October issue of *Discovery* magazine, combining a clear explanation of the advantages of LPS lights, the disadvantages of HPS lights, and biting sarcasm: "But now, the astronomy done on Palomar Mountain is seriously threatened by, as far as one can tell, concern by affluent San Diegans about their sallow nocturnal complexions. . . . So we are faced with the remarkable spectacle of the city council of a major modern metropolis, in possession of all the facts, deciding to hobble the largest productive telescope on Earth so they can illuminate smaller areas at night and pay a larger electricity bill. What can they be thinking of?. . . . There is no question that complexions appear a bit sallow with LPS lamps. Capped teeth may even look black. But so far as I know, there is not a great deal of nocturnal street carousing in those affluent San Diego suburbs that have provided much of Councilman Mitchell's support."[21]

SAN DIEGO STRIKES BACK

While LPS supporters waged their campaign on national and international levels, Mitchell initially attempted to portray the question in terms of what was best for San Diego, rather than as an attack on Palomar. He told the *San Diego Union,* "It isn't that we are against the observatory. That's not the case at all. All we're doing is responding to our constituents who are telling us that they don't want yellow lights. . . . The scientists are making it like anybody who doesn't vote for yellow lights is dumb. And that's a very unscientific approach. They're making it sound like you're against technology if you're against the yellow lights. But that's not the case. I'm for technology. But I'm also for the people of San Diego."

In contrast were his statements in a *Los Angeles Times* interview: "The low pressure lights make people look like cadavers. Their pukey yellow cast depresses a lot of people. It's going to turn our Technicolor city into black and white so they can play with the stars. . . . I don't want those damn things in San Diego, ever. Not for any reason." Perhaps Mitchell's most famous comment about LPS lights was published in another interview: "They're depressing. I talked with a psychologist and he said pink has a soothing effect. They use pink lights in insane asylums."[22]

By the end of 1983 Mitchell and his supporters had begun to stress the theme of Palomar as an outsider, which proved to have a powerful appeal to a particular San Diego mentality. One irate letter writer said, "I find it unnecessary to sacrifice the safety of [HPS] lights in a city of nearly 1 million people for an installation of any kind 50 miles away."[23]

The struggle had now created a split in the local media. San Diego was still a two-newspaper town in 1983–1984. The *San Diego Union,* which was the morning paper, supported LPS lights with numerous editorials. One said, in part, "To . . . insist on high-pressure sodium lights would be to confirm a Luddite mentality in San Diego that would merit the widespread contempt and condemnation it would surely receive."[24] The *San Diego Tribune,* the afternoon paper, supported Mitchell. This was apparent in several attacks on Carl Sagan. A November 11, 1983, column referred to him as "a pretty face, a shallow darling of the campus lecture circuit, a parrot of ideas."[25]

Mitchell also attacked Sagan, insisting that Sagan had said that the Hubble space telescope would make ground-based telescopes obsolete. Sagan repeatedly said he was being misquoted and wrote Mayor Hedgecock to deny Mitchell's claim.[26] Despite Sagan's repeated denials, Mitchell and his supporters continued to use the bogus statement.

FINAL MOVES

By late January 1984 the LPS supporters had gained the advantage. Councilman William Jones appointed a ten-member task force to advise him on the lighting issue.[27] Gloria McColl also indicated she was leaning toward supporting LPS lights. Most significant, Mayor Roger Hedgecock was indicating privately that if it came down to a tie, he would vote for LPS lights.[28]

As the vote neared, Mitchell repeatedly attacked the integrity of the city manager's office. He accused city manager Ray Blair of constantly "coming up with different figures" on LPS costs, adding, "It makes me wonder who's been getting to them." The source of his anger was a city manager's report indicating LPS lights would provide more than $209,000 in energy and maintenance savings per year over HPS lights.[29]

Mitchell's final effort was letters to the *Union* and the *Tribune,* calling on his supporters to turn out. He wrote, "If the general public involves itself more with this important issue, it is my belief that special-interest pressure will not prevail. If the residents of District 1 will turn out at the council hearing and express your desires, the destiny of our neighborhoods will follow your direction."[30]

The two newspapers also gave their final editorial judgments. The *Union* published a long editorial in support of LPS lights. The *Tribune*'s editorial was much shorter and continued the paper's support of Mitchell, referring to LPS bulbs as bug lights. The writer also attacked the efforts of LPS supporters, saying, "The facile arguments of pop scientist Carl Sagan, calling San Diego city officials provincial and small-town for their stand against 'bug lights,' hurt the cause of the . . . Palomar Observatory and the . . . observatory on Mount Laguna." The editorial concluded, however, "But, emotionalism aside, there is the indisputable fact that only the low-pressure lights will protect the work of the astronomers. The City Council can show the region and the nation that it considers that work of lasting importance by reversing its decision against the low-pressure lights."[31]

VICTORY

The city council meeting was finally held on Monday, February 6, 1984, at 2:00 p.m. The audience filled the meeting room and was divided between Palomar supporters and opponents. One or more busloads of people from Rancho Bernardo had come down to back Mitchell. The San Diego State University astronomy department also sent a contingent of professors and students. The various speakers had their say, yet again.

The most telling examples of San Diego's resentment toward Palomar were demands by several of Mitchell's supporters that the 200-inch telescope be moved. One speaker attacked Caltech for retaining attorney Paul Peterson, rather than spending the money on moving the telescope. The issue of Palomar as outsider had, by now, superseded the earlier issues of crime and economics; the speakers resented the fact that the telescope existed.

When all the speakers were done, it was Mitchell's turn. He said he wanted to help the observatory, but all his attempts at compromise had been rejected. He accused Palomar of using political pressure in hiring Peterson and attacked all those who had opposed him. When he was finally finished, Councilman Bill Cleator, a supporter of LPS lights from the start, replied in a calm voice, "Bill, you sort of remind me of the flat-worlders during the fifteenth century. I think that Columbus would still be sailing around in his own bay if he had that attitude." There was a moment of stunned silence; no one in the audience could believe what they had just heard. Finally, the audience began to clap loudly.

Next to speak was Councilman William Jones. He said that he was impressed with the $250,000 cost savings of LPS lights and the larger area they covered. He added, "If there is a chance for error, frankly, I would rather be wrong on the side of our future, wrong on the side of our observatory." He concluded by saying he believed he was wrong about his earlier support of HPS lights. Gloria McColl also said she was changing her position: "I really think we're talking about two kinds of yellow lights." It had been Jones and McColl who had provided Mitchell with victory the previous June.

Mitchell then launched into a tirade, saying that his side had not had a fair hearing the first time. Mayor Hedgecock said Mitchell had the floor and to go ahead. Mitchell lashed out, impugning the integrity of the city manager, the deputy city manager, and the employee who had prepared the report on the cost savings. Mitchell then asked who would pay for converting the existing HPS lights: Caltech, Palomar, or Mount Laguna Observatory. There was a gasp from the crowd; they realized Mitchell was now totally out of control.

City manager Blair calmly replied that the money would come out of the general fund. Mitchell repeated the question: "You mean the taxpayers?" Blair again said it would come from the general fund. Mayor Hedgecock finally cut Mitchell off with a "Yes, Mr. Mitchell, the taxpayers." This ended the debate, and the vote was held on accepting or rejecting the bids for the HPS lights.

The tally was 6 to 3 for LPS lights. The six were Gloria McColl, William Jones, Bill Cleator, Ed Struiksma, Mike Gotch, and Uvaldo Martinez. Backing the HPS bids were Mitchell, Hedgecock, and Dick

Murphy. A second vote was held, with the same outcome, to solicit bids for an LPS system and to investigate buying San Diego Gas and Electric's streetlights.[32]

3043 SAN DIEGO

In honor of the city council's action in approving LPS lights, an asteroid was named for the city. It had been discovered on September 20, 1982, by Eleanor Helin using the 48-inch Schmidt telescope on Palomar Mountain. Its provisional designation was 1982 SA. James Gibson made additional observations that allowed the asteroid's orbit to be calculated, placing it in a nearly circular orbit just outside the orbit of Mars, on the inner edge of the asteroid belt.[33]

By early 1984, with its orbit calculated, the asteroid was eligible for naming. Helin decided to name it for the city. On May 3, 1984, a plaque with a photo of the asteroid's trail was presented to deputy mayor Mike Gotch by Caltech president Marvin L. Goldberger and Helin. The asteroid is about three miles across, roughly the same size as downtown, and has a distinctive sandy composition. The naming citation read: "3043 San Diego: Named as a celestial tribute to the city of San Diego in appreciation of the city's responsiveness and cooperation in the campaign to restore dark skies for astronomers probing the universe."[34]

The year 1984 also saw advances on several fronts. While attention was on the city of San Diego, Caltech was also working for LPS lighting in the county of San Diego. In contrast to the bitter struggle waged in the city, the county accepted LPS lights with virtually no problems. Escondido, Ramona, and several other cities converted to LPS lights.[35] The county supervisors approved the conversion to LPS lights in May 1984, followed by restrictions on outdoor lighting from parking lots, billboards, searchlights, and decorative lighting in December 1984.[36]

The LPS vote marked the end of Mitchell's political career; he was defeated for reelection in November 1985 and vanished from the political scene. Hedgecock was accused of taking illegal campaign contributions. His first trial ended with a hung jury. In a second trial Hedgecock was convicted of one felony count of conspiracy and twelve counts of perjury, but it was reversed on appeal. Before a third trial was held, Hedgecock pleaded guilty to the single felony count of conspiracy, in exchange for no jail time and a $5,000 fine.[37] By this time, Hedgecock was a right-wing radio talk show host, railing against city government, liberals, Mexico, illegal immigrants, and NAFTA. Hedgecock remains one of the most politically influential people in San Diego.

Despite their failure, the HPS supporters became even more strident in their attacks on Palomar and astronomy. One letter writer lambasted the *San Diego Tribune* for siding "with pop scientist Carl Sagan and a bevy of Caltech astronomers and against . . . the citizens of San Diego."[38] Another letter called for a vote to decide "whether the majority prefers that Palomar relocate its telescope."[39]

Another attack followed the approval of county lighting regulations. One letter writer lashed out at editorial support for the rules as "half-baked nonsense and consistent twaddle." The new rules were described as "against proper and safe lighting for the taxpayers in the interest of a mere handful of job-holding astronomers." This individual further claimed that the 200-inch telescope "is not at the cutting edge of science as some weirdos seem to still believe. There is no reason whatever to inconvenience literally millions of people in Southern California so some fool astronomer can play with this particular toy."[40]

This was the year of George Orwell's *1984*. In the novel's imaginary superstate of Oceania, the Party uses the artificial language of Newspeak to make it impossible to think unapproved ideas. Orwell wrote, "It means a loyal willingness to say that black is white when Party discipline demands this. But it means also the ability to *believe* that black is white, and more to *know* that black is white, and to forget that one has ever believed the contrary."[41] In Oceania, black is white because the Party decrees it. In San Diego, the city council was about to decree that pink is white.

PINK IS WHITE, 1990–1997

The most influential group now opposed to LPS lights was the Center City Development Corporation, which continued the struggle to set up a district of restaurants, shops, and theaters. It continued to blame the lights for the problems and to seek an exception from the city council. The question was put in terms of "them versus us." Pam Hamilton, a vice president of the group said, "I guess my problem on this is who is this town for, the astronomers or the people who want quality of life and security?" The backer of the exemption was Councilman Ron Roberts. His complaint was that LPS lights did not give "the sense of security," and he thought the city should change its policy in certain limited areas. Each time, however, Roberts was on the losing end of lopsided votes.

The most serious attempt was made in March and April 1990. The council decided that additional lights were needed in high-crime areas. The opponents of LPS lights saw it as an opportunity to reopen the whole lighting issue. The attempt gained momentum when Councilwoman Abbe Wolfsheimer expressed support. She said, "We have to balance the interests of the observatories with the needs of our people. And I think the scales tip heavily in favor of our people. I think we are going to have to use the [high-pressure] lamps and move out of the low-pressure sodium."[42]

At the April 2, 1990, meeting, a representative of the Center City Association called the lighting question "an issue of safety." He complained that while astronomers talked about the "vulnerability of telescopes," they were ignoring "the vulnerability of our citizens of San Diego." He continued, "Safety has to be paramount." The astronomers argued that the difference in color between LPS and HPS had no effect on crime. The HPS supporters did not contest that point but rather argued that they *seemed* safer. A downtown bar owner told the council, "The [LPS] lights are perceived to be dimmer. The streets are perceived to be grimmer."

In retrospect, it was clear that the April 1990 effort by the HPS supporters was poorly organized. When it was the council members' turn to speak, one asked pointedly why they were dealing with this issue again. The final vote was 6 to 2 to install additional LPS lights in the high-crime areas. The two opposing were Ron Roberts and Abbe Wolfsheimer. Even limited conversion still had no support on the council.[43]

It is also now clear that the 1990 attempt marked a change in tactics by HPS supporters. Mitchell had argued that HPS would actually prevent crime. The HPS supporters now subtly shifted their position. Their answer to a growing fear of crime was to use lights that made people *feel* safer.

As the city council was revisiting the streetlighting issue, San Diego was entering a long period of decline. The two central elements of the local economy were the Navy and the local aircraft industry. With the end of the Cold War, the Navy was shrinking in size and the U.S. aerospace industry underwent a nationwide collapse. The local standard of living was falling below the state and national averages. Fear grew—fear of the future, fear of the outside world, and fear of crime. On February 20, 1992, San Diego police chief Bob Burgreen reported a 14.8-percent increase in violent crime. It was a perfect environment for the new feel-good politics of the HPS supporters.

The leader of the new crusade was Councilman John Hartley. He was a rarity in San Diego politics, a liberal Democrat in a town long known for being conservative and Republican. He also had ambitions to be mayor. Hartley had missed the 1990 vote on streetlight conversion, but in 1992 he took to the cause with the same fervor as Bill Mitchell a decade before.

Hartley raised the issue in March 1992 with a fierce attack on LPS lights, Palomar, and astronomy. "Science is not supposed to hold us hostage and take us back to the dark ages," Hartley told the *San Diego Union-Tribune*. After suggesting astronomers start looking for other sites for their telescopes, Hartley added, "Science should be able to take care of itself. I think San Diego should come first . . . and if that means that that impacts Mount Palomar so be it."

Nick Johnson, an aide for Hartley, was even more direct. In an interview Johnson suggested that Palomar Observatory had outlived its usefulness and "would make a good restaurant." He continued, "If [HPS lights] would make people feel safer, I would say, 'Nice restaurant. Palomar under the stars.'"[44]

In the newspaper interview Hartley had said, "I think people feel safer at night with white light." He added that LPS lights were "the equivalent of having no lights at all" and that the issue was "white lights versus the yellow, or bug lights." HPS lights are pink; even Mitchell always referred to them as pink, and that was the basis of his memorable line about pink lights in insane asylums. HPS supporters began using the Newspeak term "white" to describe them. Hartley stressed how HPS would make people feel safer. He could not point to any studies indicating reduced crime, but he said that the perception of safety justified replacement of the LPS lights.

The reality of streetlighting and the crime and budget situation in San Diego was far different. The city manager's report stated, "There has been no measurable relationship between streetlighting and crime or traffic accidents." Indeed, the report found that the highest crime area in the city, the intersection of Fifth and F Street, was already so well lit that no more lights could be added, nor could the wattage be increased. The same was true of three of the five highest crime areas.

San Diego police spokesman Bill Robinson said that he could not recall a single instance of an investigation being jeopardized by the LPS lights. Both Chief Bob Burgreen and Sheriff Jim Roache stated that the biggest problem was lack of jail space. Burgreen said, "Low-level street crime crooks don't spend any time in jail when we catch them because we don't have any jail space in the county."[45]

Then there was the conversion cost of between $4 million and $8 million, and $475,000 per year in added energy costs, at a time when San Diego was facing a shortfall of $25 million to $30 million. The ratio of police officers to population fell for a second year. City services, such as landscaping in city parks, were cut back. The cost did not bother Hartley, who said, "That's the right of the people." LPS supporters noted that while there was no money for added police officers or jail space, there was money for the symbolic act of replacing the LPS lights.

Hartley mixed his attacks on Palomar Observatory with offers of a compromise. He said he might be willing, "in the spirit of being a good neighbor," to "allow high-sodium white lighting" in areas south of Interstate 8, while keeping LPS lights in the northern parts of San Diego.[46] Hartley told San Diego State University officials that he did not want to harm the observatory, but the LPS lights had to be replaced. He was using salami tactics: the lighting ordinance was so broad that essentially any area could be converted.

Hartley could count on the support of three other council members, Ron Roberts, Abbe Wolfsheimer, and George Stevens. Councilman Stevens was a black minister who had been involved in civil rights since the 1960s and who had a fiery temper and a foul mouth. In 1992 he called another councilman a "white racist redneck" and threatened to "kick your ass."[47]

A feeling of doom haunted LPS supporters at the meeting of the Public Services and Safety Committee on March 18, 1992. According to Clark Dawson, vice president of the San Diego Astronomy Association, the Palomar presentation was very poor, and the LPS supporters made dull, unemotional statements. The exception was Michael Anderson. He lambasted the committee members, asking them if they realized how many policemen they could put on the street for $8 million and commenting that some small-town departments operated on that level of funding. The council members squirmed for a moment, but that passed. Dawson continued, "We stuck to the issue and tried to make our case based on reason. Unfortunately, no one wanted to be reasonable."

In contrast to the LPS supporters, Hartley packed the audience with his own campaign workers, who made one emotional plea after another that the lights be changed so San Diego would be crime-free.[48] After two hours of public statements, the council members gave their positions. Hartley said, "Now is the time to move in the direction of lights we feel safe with." He continued, "We don't have to apologize for this. We don't have to apologize to anyone." The motion passed and was sent on to the full city council.[49]

In the wake of the disastrous meeting, and with the vote scheduled for April 27, LPS supporters took action. The *San Diego Union-Tribune* opposed Hartley's efforts, noting the cost and that in many high-crime areas there was only a single streetlight per block.[50] *Science* magazine quoted Hartley's aide about Palomar as a restaurant, then sarcastically noted, "Too bad the name 'Stars' is already taken."[51] There was also a groundswell of letter writing. The date for the vote was delayed to June 1, and by early May some 1,700 letters had come into city hall from astronomers around the world.[52]

The HPS supporters were also active. One letter to the editor said, "The safety of the people of this city should come ahead of whether the observers at the Palomar Observatory or Mount Laguna can see Venus on a cloudy night." He concluded, "There are some nice mountaintops in Imperial County where they may want to relocate, far from city lights."[53]

The council vote ended in defeat for Hartley and his allies, but it was not a victory for LPS supporters, either. Hartley had already announced, "There will come a day when we will have white lights." His office had previously said that the issue would again be raised after the first of the year if significant areas were not converted. The November 1992 elections would also see a new mayor and possibly one or two new council members.

When Hartley again raised the lighting issue in early 1993, the decision had already been made. HPS lights would be installed; all that remained was to go through the motions. The Public Services and Safety Committee met on March 31, 1993, and approved the $1.64 million conversion program. The proposal was sent on to the full city council, where it was assured of victory. In the meantime, the same individual who lambasted the county lighting restrictions in 1984 called the *San Diego Union-Tribune*'s support of Palomar "asinine." He declared, "The 'important' work being done by astronomers at the observatory on Palomar Mountain doesn't amount to a hill of beans in actual fact. Little that is done in astronomy anywhere is of any practical importance."[54]

Behind the scenes, the city council's effort bordered on farce. The first indication came the very next day after the Public Services and Safety Committee approved the conversion. The city asked police and firemen to take a pay cut due to a lack of money. The level of the city council's scientific knowledge was made clear during one meeting. Dr. Robert Brucato of Palomar was, yet again, explaining that HPS lights spread light across the white-light spectrum. LPS, he explained, was concentrated in a single wavelength, which could be blocked with a simple filter. Councilwoman Wolfsheimer then suggested that astronomers make a filter to block out the whole white-light spectrum and leave the light from astronomical objects. Brucato explained that it was all white light, that there was no difference between the light from a streetlight and that from a supernova. Wolfsheimer responded, "Astronomers are smart people, and I'm sure you can figure out a way to do this somehow."[55]

The use of Newspeak was not limited to calling pink white. Ron Roberts's office mailed a form letter that began: "Like you, I am also concerned about the effects light pollution has on the viewing at Mount Palomar Observatory. That is why I support converting to white street lights, rather than increasing the number and wattages of existing yellow street lights." The letter claimed that additional LPS lights "will contribute more light pollution." Converting to HPS, Roberts said, would require fewer lights, which, he seemed to imply, would reduce light pollution.[56] (In Oceania, when the chocolate ration was reduced, it was called an increase.) Copies of the letter were sent to interested parties in Washington, D.C. Their reaction was uncomplimentary.

As the vote neared, spontaneous demonstrations were organized. One protest, by Stevens's supporters, featured people carrying hand-lettered signs and marching in a circle for the television cameras while chanting, "Ho Ho, Ha Ha, make the bug lights go away." Hartley's supporters at the final vote on September 28, 1993, carried identical, professionally printed protest signs.

The final vote was 5 to 4 for the HPS conversion in high-crime areas. A second proposal, to convert the downtown redevelopment district, passed 7 to 2. Hartley was exuberant after the vote, saying, "I think it's a dramatic victory, really for sanity. I think it's obvious bright lights are what people want." Hartley continued, "At the same time, I think it's good government. It meets the needs of the neighborhood, and of our neighbors at Mount Palomar and Mount Laguna." Stevens was more forthright, "I've often said that if I had the opportunity . . . I would have a floodlight going down every street." Geoffrey Burbidge, a theoretical astrophysicist, said of San Diego's good-neighbor policy, "It isn't sensible, it isn't logical, and it isn't right—but that isn't the way politics is done, is it?"[57]

The only positive aspect was the establishment of a study committee to develop lighting policy. David Crawford of the International Dark-Sky Association said that this should lead to a rational rather than an emotional approach to lighting issues.[58] Over the next several months, the study group met. It was broad-based, with astronomers, lighting engineers, and the public. They thrashed out a series of workable recommendations on such issues as light pollution and trespass, which were submitted to the city. The members received a thank-you letter, but the recommendations were never acted on.

The participants met differing fates. John Hartley finished his second council term and began making plans to run for mayor in 1996. The effort vanished without a trace. Abbe Wolfsheimer also completed her term and, like Hartley, disappeared into political obscurity. Ron Roberts was elected to the county board of supervisors. George Stevens was reelected to the city council. In 1996 he proposed a program called Speak! San Diego. Stevens explained, "The program was designed to get people speaking to each other, creating a more friendly atmosphere," as well as "for the purpose of reversing the tide of negative expression."[59] A year later Stevens called the other members of the city council liars and racists during a debate over a new city manager.

As astronomers had feared, the initial vote was only the first action. In February 1994 the area of Park Boulevard was added to the conversion effort. In January 1995 the conversion area expanded to all streets south of the Interstate 8 freeway in high-crime areas more than 40 feet wide, costing an additional $672,000 and raising the total conversion bill to $2.47 million.[60] In the summer of 1997 HPS lights were installed in Balboa Park. San Diego officials promised that the new HPS lights would have shields to reduce light pollution, but many of the fixtures in the downtown redevelopment area, along Market Street, and in Balboa Park have teardrop or spherical glass fixtures. They have no shielding, and half the light goes directly into the sky.

As of this writing, asteroid 3043 remains named for a small California town, "as a celestial tribute . . . in appreciation of the city's responsiveness and cooperation in the campaign to restore dark skies for astronomers probing the universe."

9

IMPACT

If the radiance of a thousand suns
Were to burst at once into the sky,
That would be like the splendor of the Mighty One . . .
I am become Death,
The shatterer of worlds.

Bhagavad Gita, recalled by J. Robert Oppenheimer, as he saw the first atomic bomb test, July 16, 1945, Trinity Site, New Mexico

IN LATE 1609 AND early 1610 Galileo turned his telescope toward the Moon and discovered that it was not a smooth body, as astronomers had believed, but rather was "uneven, rough, and full of cavities and prominences, being not unlike the face of the Earth." Galileo also observed that there were numerous "small spots" scattered over the surface; these circular features, or "craters," seemed unique to the Moon.[1]

As telescopes were improved, the scale and number of the craters became apparent. They ranged in size from the 331 kilometers of the crater Bailly down to tiny pits just visible through a telescope. The form of the individual craters also differed. Some were circular rims surrounding flat floors, while others, such as Tycho and Copernicus, were bowl-shaped with raised rims, rough floors, and a central peak.

THE EARLY HISTORY OF IMPACT THEORY

Debate began almost immediately as to their origin. Some proposed that craters were the result of explosive volcanic eruptions; others thought they were the result of impacts. In 1665 Robert Hooke, an English naturalist and observer of the Moon, tried to duplicate both possibilities. To simulate volcanic activity, he boiled a mixture of plaster of Paris. He said of the results, "The whole surface, especially that where some of the last Bubbles have risen, will appear all over covered with small pits, exactly shap'd like those of the Moon."

His attempts to duplicate lunar craters with impacts were also successful, but Hooke doubted the results were meaningful. To simulate the Moon's surface, he used "a very soft and well-temper'd mixture of Tobacco-pipe clay and Water, into which, if I let fall any heavy body, as a Bullet, it would throw up the mixture round the place, which for a while would make a representation, not unlike those of the Moon." He could not, however, "imagine whence those bodies should come; and next, how the substance of the Moon should be so soft."[2]

The debate over the origin of the Moon's craters continued for another three centuries after Hooke conducted his experiments. One explanation suggested that lava on the Moon, rather than erupting from a volcano as on Earth, created a dome or bubble. As the flow ebbed, the pressure dropped, and the dome collapsed. As the cycle repeated, the outer rim would be built up. When the cycles finally stopped, the floor cooled, forming hills, terraces, smaller craters, a central peak, or a flat plain.[3]

The impact theory was again suggested by a Munich astronomer named Franz von Paula Gruithuisen in 1828. Gruithuisen was noted for having some fanciful ideas about the Moon, and he does not seem to have been taken seriously. The first recognized astronomer to propose it was R. A. Proctor in 1873. Again the proposal did not receive significant attention, and throughout the first half of the twentieth century the volcanic theory predominated.[4]

The modern rebirth of the impact theory came with the publication of Ralph B. Baldwin's book *The Face of the Moon* in 1948. Baldwin, an astrophysicist, combined observational data, theoretical studies, and geological information to support his impact ideas. Moon studies were still far from mainstream astronomy—Baldwin's day job was as vice president of the Oliver Machinery Company.[5]

By the late 1950s and early 1960s the majority of American astronomers had come to believe that the Moon's craters were due to impacts. A few astronomers still backed the volcanic theory, however. In America J. E. Spurr argued for the volcanic theory in a series of books published between 1944 and 1949. It was the USSR, however, where the volcanic theory found the most support. On the night of November 3, 1958, the Soviet astronomer Nikolai A. Kozyrev was taking a spectrum of the crater Alphonsus when he saw a reddish glow in the crater's center. He took another spectrum that indicated the presence of molecular carbon. Kozyrev believed he had observed an eruption on the Moon.[6] Professor Alexander Mikhailov, chairman of the Astronomy Council of the Soviet Academy of Sciences, said, "We can now regard as completely unfounded the existing view of the origins of [lunar craters], which ascribes them to the fall of meteorites."[7]

The problem was that as long as astronomers were limited to what they could see, the debate could never be settled. The "impacters" asked how a small body like the Moon could undergo volcanic activity on a size and scale never seen on Earth. Another objection was the lack of any craters showing intermediate steps, such as a huge dome.[8] The volcanists responded that the lower gravity on the Moon, one-sixth that of Earth, would allow larger formations. They also argued that when two craters overlapped, it was the smaller crater that broke the larger one. This indicated the smaller crater had formed later, which was taken as evidence of diminishing volcanic activity.[9]

As the debate continued, there was a revolution in the understanding of Earth's geology. During the seventeenth century, geologists were influenced by biblical chronology, such as that by Bishop James Ussher, which stated that the world had been created in 4004 B.C. It was presumed that geological changes had to be very rapid, a doctrine called catastrophism. As geological knowledge increased during the eighteenth and nineteenth centuries, catastrophism became untenable. In 1830 Charles Lyell developed the concept of uniformitarianism, which held that the slow processes seen today, such as erosion and the deposition of sediment over billions of years, are the key to the geological record.[10]

Even as uniformitarianism was being formulated, there was additional speculation that comets had impacted Earth. Edmond Halley was the first to suggest that a comet might strike and could affect Earth's climate. In 1688 he suggested that the biblical flood might have been due to a comet impact. Halley expanded on this in an address to the Royal Society of London on December 12, 1694, pointing out that a comet impact might well cause worldwide extinction. In 1750 Pierre-Louis Moreau de Maupertuis wrote that comets could hit other planets and that the resulting heat and contamination of the air and water would result in mass extinctions. He added that the confused strata of Earth suggested such impacts had already occurred.

More than seventy years later, cometary impacts were a recurring theme in the works of Lord Byron. He believed that such impacts had occurred many times in the past, wiping out the previous inhabitants of Earth. Lord Byron also felt that humans were perhaps only temporarily in the ascendent and might also fall to a comet.[11]

Such speculation and poetical musings ran afoul of the flip side of uniformitarianism. In declaring that geological features were due to slow changes, it also rejected any possibility that catastrophes had ever played a role in Earth's history. What Halley, Maupertuis, and Lord Byron suggested was seen by professional geologists as catastrophism, which was unacceptable. The stage was now set for a geological dispute that lasted more than a century. It began with a single unique geological formation originally called Coon Butte. It would lead to far-off worlds and to a day, long ago, when the world ended.

COON BUTTE

As white settlers began arriving in what is now the state of Arizona, they found a remarkable feature. From the flat plain, it appeared to be a low butte. The slope was actually a raised rim between 148 and 223 feet high. It surrounded a steep-sided pit about 4,000 feet in diameter and 613 feet deep.[12] Coon Butte first came to the attention of geologists in the 1880s when local sheep herders found what proved to be nickel-iron nearby. A few years later a prospector said that he had found a deposit of iron so rich that it could be picked up from the ground. Samples of the iron eventually reached A. E. Foote, who recognized them as being of meteoritic origin.

In 1891 Foote traveled to Coon Butte to investigate. He described the deep crater, raised rim, and uplifted layers of sandstone and limestone. He could not find any volcanic rock, however, and was "unable to explain the cause of this remarkable geological phenomenon." Foote did find significant amounts of meteoritic iron, and his discoveries attracted the attention of America's most respected geologist, G. K. Gilbert.

At the time Gilbert was the chief geologist of the U.S. Geological Survey. He arranged for a colleague, W. D. Johnson, to undertake a detailed survey of the crater. Johnson reported that despite the lack of volcanic rock, he thought it had been formed by a steam explosion. This conclusion, however, did not explain the meteoritic iron. Gilbert was still curious and decided to see the crater for himself. He wrote to a friend, "The errand is a particular one. I am going to hunt a star."

In November 1891 Gilbert and Marcus Baker, of the U.S. Coast and Geodetic Survey, visited Coon Butte. Gilbert assumed that because the crater was round, the meteor must have hit vertically, in which case, a huge amount of iron would be located directly beneath the crater floor. But when Gilbert made a careful magnetic survey, there were no indications of any such mass. He also calculated that the volume of material in the rim would just about fill the crater. This, Gilbert thought, also indicated that there was no buried iron mass, as material equal to its volume should also be in the rim. Based on these assumptions, Gilbert decided that Coon Butte was not formed by a meteor impact. He wrote to his friend, "I did not find the star, because she is not there." He concluded, like Johnson, that the crater was the result of a steam explosion.

Like Hooke before him, Gilbert also conducted experiments with impacts in different materials. He dropped marbles into porridge and lumps of mud into beds of mud, fired pellets of clay with a slingshot into a bed of clay, and shot various materials from a large-bore musket into beds of the same material. By 1892 the results had convinced him that although Coon Butte was formed by a steam explosion, the Moon's craters were formed by impacts. There was still a loose end. In his tests he found that when the impact was at an angle, the crater was usually elliptical. The craters of the Moon were always round.

In 1895 Gilbert gave a lecture to the Geological Society of Washington titled "The Origin of Hypotheses, Illustrated by the Discussion of a Topographic Problem." Gilbert used the example of Coon Butte to describe the process by which a scientific hypothesis was developed, then tested, and either confirmed or discarded. He went over the various hypotheses for the crater and concluded that it was caused by a steam explosion. He was still troubled that a large meteor should fall exactly at the spot where a steam explosion occurred.

Gilbert closed his lecture by noting how a single question by a correspondent had the "ability to unsettle a conclusion which was beginning to feel itself secure." The reporter asked whether some of the rock might have been compressed by the force of the impact so it occupied less volume. If true, then one of the arguments supporting the steam explosion hypothesis would be invalid. Gilbert concluded that such a question underlines the tentative nature of any theory. These "are ever subject to the limitations imposed by imperfect observation. . . . In the domain of the world's knowledge there is no infallibility."

Despite Gilbert's nondogmatic comments about the crater's origin, all most people heard was his acceptance of the steam explosion theory. His stature was such that this was considered the final word on both Coon Butte and the subject of meteor impacts on Earth. The message of his lecture was lost, and any other possibilities were discounted. Coon Butte itself was dismissed as a "relatively trifling and local affair."

One man who disagreed was Daniel M. Barringer, a successful mining engineer and lawyer. In 1902 he learned of the crater and discussed it with a friend, Benjamin C. Tilghman, a mathematician, physicist, and expert on the impact effects of heavy artillery. Barringer and Tilghman examined some of the meteor samples and became convinced that the crater had been formed by an impact. Despite Gilbert's failure to detect any magnetic anomalies, they also believed that a large mass still remained beneath the crater, containing deposits of iron, gold, platinum, and iridium. Sight unseen, they acquired mineral rights to the crater and organized the Standard Iron Company.

Over the next several years Barringer and Tilghman made extensive surveys, collecting evidence that it was a meteor crater. This included mapping the scattering of rocks around the crater, including boulders weighing some 5,000 tons found a mile from the rim, and vast quantities of powdered rock inside the crater and rim. Both pointed to an explosion far greater than could be attributed to steam. They also failed to find any volcanic rock or any evidence of hot spring activity. The most important evidence was the composition of the rim: a mixture of meteor iron, boulders, and powdered rock. This, in their view, removed any possibility of a steam explosion. If the meteor had fallen before the crater was formed, then it would be covered by the debris blasted out of the crater. If the meteor fall had occurred after, then the iron would be on top of the debris. The only way a mixture could occur was if the meteor had formed the crater.

Barringer, like Gilbert, assumed that because the crater was round, the meteor must have landed vertically. He drilled in the floor of the crater;

the core samples showed hundreds of feet of powdered rock, followed by zones of crushed rock, and finally intact sandstone. There was no evidence of volcanic activity, but no large iron mass was discovered either.

Barringer and Tilghman published their results in separate papers in 1905. Gilbert's opinion still held sway, however, and Barringer and Tilghman were dismissed as upstarts. Although a few geologists and astronomers now accepted an impact origin for the crater, for most the idea was still considered too fantastic to believe. The attempt to find the iron mass was also running into trouble. By July 1908 some 28 holes had been drilled into the crater floor, but the iron mass still could not be located. By this time Tilghman had sunk several thousand dollars into the project. He lost confidence in the prospects of success and dropped out in 1909. Barringer carried on alone.[13]

Failing to find the iron mass, Barringer reconsidered the problem of the impact angle. As an experiment, he fired rifle bullets into the ground and discovered that even an oblique impact would result in a round crater. Barringer decided that the meteor had come in from the north at a 45-degree angle. The south rim was raised above that of the north side, and Barringer decided that the iron mass must be below it.

Barringer spent the next decade trying to convince skeptical scientists that the crater was formed by a meteor and seeking capital to resume the search for the iron mass. Finally, in 1922 Barringer was able to arrange for new drilling. As before, this attempt, and another two years later, ended in failure. In all, Barringer had spent more than $600,000 trying to find the iron mass. Only the abandoned drilling equipment on the crater floor remained. Barringer's quest had ended. He died in 1929.[14]

It was now apparent there was no iron mass to find. George P. Merrill, curator of the Department of Geology at the U.S. National Museum, had earlier suggested that the energy of the impact was so great that the iron meteorite was vaporized. By the late 1920s, after repeated unsuccessful searches, his view was gaining acceptance. Before any more money was spent, a resolution of the question was needed. F. R. Moulton, a mathematician and astronomer at the University of Chicago, was asked for advice. Moulton had also done ballistic research during World War I. He determined not only that the meteor was much smaller than had been believed but also that it had been destroyed on impact.[15]

Yet, Barringer's efforts were not in vain. A slow change in scientific opinion had already begun. In 1928 Merrill reported the remarkable findings of a Soviet geologist named Leonid A. Kulik. The story actually began twenty years before. At 7:14 a.m. on June 30, 1908, a fireball nearly as bright as the Sun was seen over the area of the Podkamennaya Tunguska River in Siberia. The fireball exploded with a force equivalent to 12.5 million tons of TNT, a blast so powerful that it rattled vodka bottles 800 kilometers away and was detected by seismic stations across Eurasia.[16]

Following the Tunguska event, as it became known, unusually colorful sunsets and sunrises were seen throughout Europe, Scandinavia, and Russia. In Central Asia and Siberia the night sky was bright enough to cast shadows and read a newspaper by. Similar light nights, caused by ice-covered dust grains high in the atmosphere, were seen in Europe.

Although press accounts appeared in July 1908, the event was forgotten for more than a decade. It was not until 1921 that Kulik heard of the event and decided to visit the Tunguska area. Witnesses told him of being knocked to the ground by the blast, of a wave of heat so intense it could be felt 60 kilometers away, and of an entire forest that had been flattened. Kulik, convinced that an enormous meteor had fallen, returned to Moscow seeking funding for an expedition to the site. It took six years. In April 1927 Kulik and two local guides trudged through the taiga and swamps, enduring clouds of mosquitoes they dubbed flying alligators. Finally they reached a mountaintop, and Kulik was astonished by what he saw. A charred expanse of felled trees extended from horizon to horizon. Only then did Kulik fully realize how great the explosion had been. One of his guides told him solemnly, "This is where the thunder and lightning fell down."[17]

Over the next decade Kulik made three more expeditions to the area. The scorched area was kidney-shaped and covered more than 2,000 square kilometers. The trees within the area were knocked down and pointing away from the center of the blast. At the center, the trees were still standing but were charred and stripped of branches. Kulik had hoped to find meteorite fragments and impact craters, but neither he nor any of the subsequent expeditions ever found any.

In 1930 English astronomer F. J. W. Whipple proposed that the Tunguska event was caused by the explosion of a small comet high in the atmosphere. Only dust would remain, explaining the lack of meteorites and craters, while the debris from the tail would explain the light nights. A comet remained the preferred explanation for the next 50 years. In the early 1980s, however, support grew for the idea that an E-type asteroid had been responsible. This was based on the way rocky meteors break up in the atmosphere. As the asteroid, estimated to be at least 80 meters across and traveling at 22 kilometers per second, plunged into the atmosphere, it encountered air resistance so great that it was brought to a complete stop in a fraction of a second. The kinetic energy was released in the form of heat, which vaporized the asteroid, leaving only fine dust.[18]

As Kulik was beginning his research at Tunguska, additional meteor craters were being recognized. By 1950 about a dozen meteor craters had been identified as having been formed in the previous 100,000 years. The Arizona crater, now called Barringer Crater or more commonly Meteor Crater, was simply the largest and best preserved. Despite all the evidence that had been amassed, many geologists still believed that it had been formed by a steam explosion.

The crater was now a tourist attraction, catering to travelers on nearby Route 66. In the fall of 1952 a young U.S. Geological Survey employee, his wife, and a colleague were returning from a field trip to the Grand Canyon when they decided to stop at Meteor Crater. They were excited to be seeing it for the first time, but when they arrived in their battered government-issue jeep station wagon, they discovered there was an admission charge. They were down to nickels and dimes and had only enough money to buy some fried rice for dinner that night. They took a quick look over the rim at sunset. On that day, Eugene Shoemaker saw his future.

EUGENE M. SHOEMAKER

Gene Shoemaker's interest in geology was first sparked by a set of marbles given to him by his mother when he was seven. A year later, while traveling with his father in the Black Hills of South Dakota, he began to collect samples of rose quartz and other minerals. After finishing the fourth grade, Shoemaker enrolled in a special science course, which convinced him that he wanted to be a geologist.

He enrolled at Caltech in the fall of 1944. The school was operating on an accelerated wartime schedule, with three semesters per year and no summer vacation. Shoemaker graduated at the age of nineteen but remained at Caltech to earn his master's degree in 1948. That same year he joined the U.S. Geological Survey and began searching for uranium deposits in western Colorado. He was staying at a small mining camp and had to drive five miles to the town of Naturita for his meals. As he drove to breakfast one morning, he had a realization—humans would reach the Moon in his lifetime. Moreover, he would do what it took to be at the head of the line when that day arrived. In 1948 the concept of spaceflight was regarded by all but a few academics as beneath their dignity. Shoemaker told no one of his dream, lest it be considered silly.

Shoemaker began studying those aspects of geology that he thought might be a road to the Moon. The first step came in the summer of 1951, while he was a Ph.D. candidate at Princeton. He was assigned by the U.S. Geological Survey to study uranium deposits in the Hopi Buttes volcanoes in northern Arizona.[19] At this point, Shoemaker said later, he still had "an agnostic view" about Meteor Crater. His mapping of the Hopi Buttes gave him an understanding of how explosive volcanoes worked. The craters they produced (called Maar-type craters) did resemble some features on the Moon and were similar in shape to Meteor Crater.

It was not until two years after his first view of Meteor Crater that Shoemaker's thoughts returned to its origins. In 1954 Dorsey Hager wrote a paper reporting that oil drillers near Meteor Crater had discovered geological features called evaporites nearby. Hager suggested that the crater had been formed by the collapse of a large salt dome. Shoemaker wrote to Hager, who sent a sample of fused quartz from the crater by return mail. The material, which had been puffed up like pumice, came from the holes drilled by Barringer years before. The material was at the center of the controversy: Hager argued the glass was volcanic, while Barringer had maintained it was sandstone melted by the impact. To settle the question, Shoemaker sent the sample out for spectrographic analysis.

The results showed that the glass had the same composition as the Coconino sandstone under the crater. To fuse the sandstone would require temperatures too high for lava. The pure quartz sandstone was very hard to melt, and it did not look like material in volcanoes. To Shoemaker, the evidence was clear—Meteor Crater could not have been formed by a volcano or a collapsed salt dome. The only source for temperatures high enough to produce the fused quartz was an asteroid impact.[20]

It was still a huge intellectual leap to go from a small meteorite lying on the ground to a mountain falling from the sky. Two years after the results of the fused quartz analysis, Shoemaker had a unique opportunity to test his impact interpretation. He was involved in a program that mapped the craters formed by the Jangle U and Teapot ESS nuclear tests, underground explosions of 1.2 and 1 kiloton bombs. The craters were 75 meters and 100 meters across. To estimate the effects of a larger blast, Shoemaker returned to Meteor Crater, where he was astonished to see the very same features he had mapped in the smaller Teapot ESS crater, only on a far larger scale. By this time he was all but sure it had been formed by an impact, but this was the final proof. Meteor Crater really was a meteor crater. His examination of the nuclear craters had allowed Shoemaker to identify the detailed structure of impact craters and the processes by which they were formed.

Shoemaker continued to search for evidence of impact craters. In the summer of 1960 he, his wife, Carolyn, and his mother went on a trip to southern Germany. Shoemaker wanted to look at Ries Basin, a fifteen-mile-wide depression attributed to a volcano. Shoemaker suspected it was from an impact, and he thought he had the means to prove it. That spring Joseph Boyd had published a paper describing the conditions that would cause a high-pressure form of silica to crystallize. Shoemaker thought that an impact would cause this mineral, called coesite, to form. More important, it could serve as a signature for an impact crater. Shoemaker's hunch was confirmed when Edward Chao found coesite in rocks from Meteor Crater.

They arrived at Ries Basin toward sunset. Shoemaker examined some of the shock-formed rock. A single look with a hand lens confirmed they were impact rocks. They spent the night at a campground, then went into the town of Nordlingen to airmail two rock samples to Chao. Eugene and Carolyn then strolled through the town, stopping at St. George Church. It was the largest geological sample he had ever seen. It was built from the local rocks, and they were filled with glass formed by shocked and melted rock. Chao soon confirmed the presence of coesite in the samples. Ries Basin had been formed by the impact of a 2-kilometer asteroid some 15 million years ago. Twenty times the size of Meteor Crater, it was proof that very large impacts had occurred in the recent geological past.[21]

Despite all his accomplishments, Eugene Shoemaker never reached the Moon in his lifetime. In 1963 he was diagnosed with Addison's disease. Although the condition could be treated, it ended any possibility that Shoemaker might become an Apollo astronaut. He was, however, chairman of the committee that selected the first group of scientist astronauts, a participant in astronaut training, and a member of the science team for the Apollo landings.[22]

The six Apollo landings ended three centuries of debate over the origin of lunar craters. The lunar samples were composed of shattered rock fragments and contained glass formed by impacts. They were identical to the samples from Meteor Crater, Ries Basin, and the other confirmed impact craters found on Earth. Impacts had formed the Moon's craters, not volcanoes.

The samples also allowed reconstruction of the Moon's impact history. Some 4 billion years ago the Moon had been subjected to an intense bombardment, ending with a series of huge impacts that created the great mare basins on the Earth-facing side. Between 3.8 and 3.1 billion years ago there was a period of volcanism; great eruptions flooded the basins, creating the lunar seas such as the Ocean of Storms. When this period ended, so did lunar volcanism; the Moon became a cold and dead body, which continued to be battered by later impacts. Shoemaker had been right.[23]

The contribution of Eugene Shoemaker to our understanding of the forces that shaped the face of both Earth and the Moon was profound. Like most ideas in science, that of impact had been long proposed. It was Shoemaker, working almost alone, who had proven it to be true. In the process, he had overturned the conventional geological wisdom, which denied that impacts occurred. Yet this was only the prologue to the culmination of his life's work in impact studies, a quarter of a century after the first Moon landing.

A THIN LAYER OF IRIDIUM

In parallel with the debate over impact craters in the 1960s was another centered on the theory of continental drift, the idea that Earth's landmasses moved horizontally relative to each other over time. When this was first proposed by German meteorologist Alfred Wegener in the 1920s, geologists ridiculed it. By the late 1960s, however, evidence had accumulated that proved continental drift was correct. The results were far-reaching; the acceptance of continental drift, or, as it became known, plate tectonics, transformed geology. The 1970s was a dynamic period, with new discoveries in every field.

The most important evidence for plate tectonics was paleomagnetism. As layers of sediment were laid down or as lava cooled, magnetic minerals within it aligned themselves with Earth's own magnetic field. These fossil compasses recorded the continent's different orientations at the times the layers were formed. The layers in sedimentary rocks could then be dated by looking for the fossil shells of ocean plankton called forams. The result was a chronological reconstruction of a continent's movements over time.

The search for paleomagnetic data brought a geologist named Walter Alvarez to the medieval Italian village of Gubbio. In the mountains above the village is an uninterrupted series of limestone layers. As Alvarez examined the layers, he noticed an unusual feature. In one layer, there were abundant forams. In the next layer, the forams were few in number and much smaller in size. Between the two was a layer of clay, about a centimeter thick, which lacked any fossils. This was the KT boundary, the separation between the end of the Cretaceous era and the start of the Tertiary era. (Geologists use "K" for Cretaceous.) The clay layer also dated to about the same time as the extinction of the dinosaurs, some 65 million years ago.[24]

By the mid-1970s the prevailing geological opinion was that the dinosaurs had evolved during a period of mild and constant climate. A shallow sea had then covered what is now the central United States, extending from Canada to the Gulf of Mexico and from the southern states up the Atlantic Seaboard. The marshes surrounding this sea were ideal for the dinosaurs. About 80 million years ago this inland sea started to retreat, accelerating about 65 million years ago and resulting in a cooler, drier climate, with less vegetation. The dinosaurs, unable to adapt, declined over several million years before finally becoming extinct. Although fast by geological standards, it fit the doctrine of uniformitarianism.[25]

Walter Alvarez was not so sure; the extinction of the forams seemed sudden, even, he dared to think, catastrophic. As the layer seemed to have been formed at the same time as the extinction of the dinosaurs, might they also have vanished catastrophically? The key question was how long the clay layer had taken to form. He turned to his father, Luis W. Alvarez, a physics professor at the University of California at Berkeley with a wide-ranging and illustrious career. During World War II Luis had developed the ground controlled approach radar to land airplanes in bad weather, worked on the atomic bomb, and flew in an observation aircraft during the bombing of Hiroshima. After the war he had participated in the CIA-sponsored Robertson Panel on UFOs, used cosmic rays to search Chephren's Pyramid for hidden chambers, and received the 1968 Nobel Prize for Physics.

Luis Alvarez suggested looking for traces of iridium from meteorite dust in the clay, a test that required very precise measurements. If the clay layer was deposited over several thousand years, there would be 0.1 parts per billion of iridium. If it took only a few years, there should be none. When the tests were completed in late June 1978, the results were unbelievable. The clay layer contained 9 parts per billion of iridium.[26]

The question became, Was this a local event or something worldwide? Walter Alvarez learned of three more KT boundary sites that showed high iridium levels, in Denmark, Spain, and a deep-sea core sample. It was now clear that the event was worldwide, and Walter and Luis Alvarez considered a wide range of possibilities, including an asteroid impact. Asteroids were rich in iridium, in contrast to Earth, but how, they asked themselves, could even a large impact cause extinction on a planetary scale?

Luis Alvarez then remembered the eruption of Krakatoa in 1883. It had blasted a huge dust cloud into the upper atmosphere, darkening the sky for more than a week. The iridium data indicated an asteroid 7 to 10 kilometers in diameter. If an asteroid that large struck Earth, it would release the energy of 100 million megatons of TNT and blast an even greater volume of dust into the atmosphere. This, they estimated, produced a blackout of two to five years, which caused the food chain to collapse, killing two-thirds of the species on Earth.[27]

Luis and Walter Alvarez announced the theory at a meeting of the American Association for the Advancement of Science on January 5, 1980. Although they were the first to offer scientific evidence that an impact had caused the extinction of the dinosaurs, the idea had been suggested before. In a 1942 issue of *Popular Astronomy* magazine, H. H. Nininger suggested that asteroid impacts caused geological boundaries and the resulting mass extinctions. He did not specifically mention the dinosaurs, however. Similar links between geological boundaries and asteroid impacts were proposed in 1958 by Ernst Opik, in 1973 by Harold Urey, and in 1979 by Victor Clube and Bill Napier.

The first explicit mention of an impact causing the extinction of the dinosaurs was in a book titled *Target Earth: The Role of Large Meteors in Earth Science*. It was privately printed in 1953 by Allan O. Kelly and Frank Dachille. The mechanism they proposed—a large impact that tilted Earth's spin axis and resulted in massive flooding—was untenable. A more significant proposal was made in 1956 by M. W. De Laubenfels of Oregon State College. In an article in the *Journal of Paleontology,* De Laubenfels scaled up the impact effects seen at Tunguska and proposed that the dinosaurs had been killed off by superheated winds. None of these were more than suggestions, and all were ignored.[28]

THE GREAT DINOSAUR EXTINCTION DEBATE

The response among paleontologists to the Alvarez impact theory was immediately and nearly unanimously negative. David Raup of the University of Chicago said later that it was considered "an appalling idea."[29] The most outspoken critics of the impact theory were William A. Clemens of the University of Berkeley and Charles B. Officer, Charles L. Drake, and Robert Jastrow, all of Dartmouth College. Much of their attack was directed at the interpretation that the iridium came from an asteroid. Officer and Drake analyzed the material in the layer and declared that it was deposited over a period of 10,000 to 100,000 years. They also concluded that the iridium had come from massive volcanic eruptions. The Deccan Traps, a massive series of lava flows in India, had been dated to this period. They argued that the volcanic gases released had caused climatic changes.[30]

Officer and Drake's argument was backed by two other researchers, Michael R. Rampino of NASA's Institute for Space Studies and Robert C. Reynolds of Dartmouth. They assumed the clay layers at each of four sites (in Denmark, Spain, Italy, and Tunisia) would be uniform in composition, reflecting the material from the asteroid and the debris blasted out of the crater. Concluding that the clay at the sites "is neither mineralogically exotic nor distinct from locally derived clays above and below the boundary," they argued that some volcanic ash was rich in iridium and that the clay itself was altered glassy volcanic ash.[31] All the sites were ocean sediment, and they argued that the clay was the result of chemical action of sea water on rapidly cooling lava.[32]

Similar objections were raised concerning the fossil record. Based on samples from Montana, Wyoming, and Canada, William Clemens argued that the extinction of plants did not coincide with that of the dinosaurs and other land animals. Rather, they were spread out over a period of 100,000 to 1 million years. When the iridium layer was formed, Clemens argued, the dinosaurs were already gone.[33] Robert Jastrow proposed the reverse, that five types of Asian dinosaurs and at least eleven types in North America survived until long after the impact.[34]

By 1983 critics of the impact theory were claiming victory. In a *Science Digest* article, Jastrow noted the excitement that the impact theory had caused in the scientific community, but added, "Science . . . should be 'the great antidote to the poison of enthusiasm.'" He recounted the critics' evidence and stated the established belief in a gradual climatic change as the cause of the extinction. Jastrow concluded by saying, "And so the problem of the dinosaur extinction has become a non-problem."

Behind the scientific arguments of Jastrow's article was an ill-concealed contempt for the supporters of the impact theory. Jastrow accused Luis Alvarez of "pulling rank on the paleontologists," adding, "Physicists sometimes . . . feel they have a monopoly on clear thinking." He argued, "A physicist's methods only work well on physics problems. . . . Perhaps the experts on ancient plant and animal life didn't know much math, but they knew their fossils, and the fossils told them Alvarez was wrong."[35] In looking back on the response, some suggested that older, established paleontologists resented finding their once sleepy field invaded by interlopers from other fields such as nuclear chemistry and astronomy and were hostile toward them. Raup also noted that paleontologists knew nothing about the research into impacts done during the 1960s and 1970s.[36]

Despite the critics' confidence, evidence favoring an impact continued to accumulate. The number of sites showing iridium anomalies increased from 4 in 1980 to 23 in 1982 and then to more than 80 by 1986.[37] More important, one of the early sites discovered was in nonmarine coal-swamp rocks in New Mexico, eliminating the possibility the iridium was from sea water. Later, additional nonmarine KT boundaries were found in New Mexico, Colorado, Wyoming, and Montana and at Canadian sites in Saskatchewan and Alberta.

The basic question concerned the source of the iridium: impact or volcano? Attempts to answer this occupied the first years of the debate. Gases from the Kilauea volcano in Hawaii were shown to contain iridium, supporting the contention that the material in the clay layer had come from volcanic eruptions. To test the idea, the ratios of iridium, ruthenium, and rhodium (three of the platinum-group elements found in meteorites) were compared. The ratios in the KT boundaries were the same as those in meteorites but completely different from the ratios found in volcanic gases.[38]

Added support came from a different test. Karl K. Turekian and Jean-Marc Luck of Yale measured the ratio in the KT boundaries between two isotopes of osmium. In Earth samples the ratio of osmium 187 to osmium 186 is 10 to 1. In meteorites it is 1 to 1. They analyzed rocks from Raton Basin in Colorado and from Denmark and found ratios of 1.29 to 1 and 1.65 to 1. Turekian said later that before doing the study, he was not convinced of the impact theory; whatever the truth of the impact theory, it was now clear to him that a large amount of meteoritic material had struck Earth 65 million years ago.[39]

A team of scientists at the Scripps Institute also developed evidence that the iridium did not come from a volcanic eruption. They found that the clay layer contained an amino acid that was very rare on Earth, nonexistent in volcanic gases, but very common in carbonaceous chondrite meteorites. As the amino acid was mixed in with the iridium, it seemed logical that both had arrived on Earth from outer space.[40]

One of the attempts to determine the source of the iridium had an unexpected result. Wendy S. Wolbach, Roy S. Lewis, and Edward Anders of the University of Chicago attempted to detect inert gases, such as xenon and neon, in the clay layer. Unable to detect any, they concluded that the object had been nearly entirely vaporized by entry and impact. In the three samples, from Denmark, Spain, and New Zealand, they did find soot particles that resembled particles from forest fires. The researchers concluded the soot was the result of wild fires on a planetary scale and estimated that the total soot was equal to 4 percent of the total land biomass existing at the time. The distribution of the soot was also nearly uniform, as the amount in the European and the New Zealand samples differed by only a factor of two. The fireball from the impact and the expanding cloud of hot rock vapor, they estimated, could start fires more than 1,000 kilometers away from the crater. The wild fires could then burn uncontrollably across an entire continent.

Up to this point, only rock dust had been considered as the extinction mechanism. The discovery of the soot changed the picture. The soot would block out virtually all the sunlight, carbon monoxide and other toxic gases could reach dangerous levels, and Earth would be cooled until most of the soot had settled out of the atmosphere.[41]

The debate about the fossil record could not be so precisely resolved. Fossils of large land animals, such as dinosaurs, are rare. Critics of the impact theory argued that the fossil record indicated a gradual decline rather than a sudden, overnight extinction. Phil Signor and Jere Lipps at the University of California, Davis, were able to show that a sudden extinction would look like a gradual one if the number of fossils was low, a principle that became known as the Signor-Lipps effect. That and the failure to find any complete dinosaur fossils or footprints above the iridium deposits cast doubts on the argument that the dinosaurs were already gone or had outlived the formation of the the clay layer.

The flip side was that as the number of fossils increased, the gap would vanish. The ideal fossils are those that are extremely common, overcoming the statistical problem of the Signor-Lipps effect. In marine strata, this condition is met by single-cell forams; in land sediments, it is plant pollen that provides a marker. At Italian and Spanish sites, the forams can be found in every rock, right up to within millimeters of the iridium layer. In land sediments from New Mexico, pollen shows a sudden extinction of some plants exactly at the layer and then a sudden profusion of ferns right after.[42]

By 1988 there had been numerous scientific papers and news articles on the question of impact extinction. Parallel to the scientific debate was another—one that was more personal and highly emotional. The issue had now become politicized. The 1980s was the period of the Reagan defense buildup and the start of the Strategic Defense Initiative (SDI) to build an antimissile system. Many scientists were vocally opposed to both and were backing a nuclear freeze on strategic forces. Public fears of a U.S.-Soviet nuclear war were the highest since the Cuban missile crisis in 1962.

The impact theory entered the nuclear debate indirectly. By proposing that the dinosaurs had died because dust and debris from an asteroid or comet impact had blocked out the Sun, reducing temperatures, the theory created the vivid image of a nuclear winter caused by the smoke from burning forests and cities. The study on soot in the iridium layer explicitly linked the impact theory with computer studies of nuclear winter.[43] At the same time, the critics' view that the extinction was caused by volcanic activity also had a political dimension. The volcanic theory proposed that the massive release of gases, such as carbon dioxide, had caused a greenhouse warming of the world's climate. There was an obvious connection between this volcanic climate change and fears about global warming caused by the burning of fossil fuels.

As the scientific, political, and personal debate continued, the accumulating evidence favored the impact theory. It came from many fields and from around the world. The only bright spot for supporters of the volcanic theory was evidence that the Deccan Traps did date from the same time as the clay layer. Opinion among paleontologists continued to be negative toward the impact theory. Jastrow quoted Norman Newell, a member of the National Academy of Sciences: "A catastrophe was not responsible for the major extinctions. Disappearance of the major forms of life was gradual over millions of years." He also cited a poll finding that 53 percent of American paleontologists believed that an asteroid impact had nothing to do with the extinction of the dinosaurs.[44]

There was still a basic flaw in the impact theory. None of the less than 100 impact craters known in 1980 matched the requirements of the theory. Most were too small. The crater had to be 150 to 200 kilometers in diameter to produce worldwide effects. Only three known craters were that large, and none was the right age. As long as the crater was not found, supporters of the volcanic theory could argue that no such impact had occurred.

THE SEARCH FOR THE CRATER

The first question was whether the impact had occurred on land or in the ocean. In a number of the iridium layers, spherules (white objects the size of sand grains) were found. When they were viewed under a microscope, they were identified as feldspar, indicating an ocean impact. There were no deep-sea sediment cores that showed signs of a crater, but the crater could be in the 20 percent of the ocean crust that had been subducted into deep ocean trenches and remelted. If this was the case, the crater itself would have been lost forever.[45]

By 1987, however, another possibility had become apparent. Bruce F. Bohor and his U.S. Geological Survey colleagues found tiny fragments of quartz, indicating a continental impact, in the clay layers at sites in Montana, Italy, and Denmark and in a core sample from the Pacific. The quartz fragments at each of the sites appeared to have come from the same geological formation, and they all showed evidence of a tremendous impact. The quartz contained multiple intersecting sets of parallel fracture planes. This required an impact in excess of 1.3 million pounds per square inch. Such shocked quartz is formed in underground nuclear tests and in meteor craters. Quartz from volcanic eruptions shows only a single set of planes. Bohor said, "I think what we're reporting makes it pretty inescapable now that there was an impact that shocked the quartz and distributed the grains worldwide. There's no evidence you can get the kind of features we are seeing from any kind of volcanic activity."

As might be expected, Charles Drake rejected his conclusions. Both agreed that volcanic eruptions could produce a single set of parallel fracture planes in quartz, but where Bohor argued that only an impact could produce the multiple intersecting planes, Drake countered that they may exist in volcanic quartz, even though no one had found any.[46] Drake's rejection of the impact origin of the quartz also represented a rejection of the research done by Shoemaker and the geological results of the Apollo flights. As a counter, all he could offer was to suggest that similar features may be produced by a volcano.

The shocked quartz also pointed to the location of the crater. The spherules seemed to indicate an impact in the ocean crust, but the quartz suggested a continental impact. Because the quartz grains from Montana were more than half a millimeter across, and those farther away were smaller, Bohor concluded that the impact may have been in North America.

Further clues came from Haiti and Texas. One of the first KT sites identified was near the village of Beloc on the southern tip of Haiti. Several years later another was discovered along the Brazos River in Texas. It was suggested that the Texas site might represent tsunami-deposited sediment from the impact, but its importance was not then realized. The first to do so was a University of Arizona graduate student named Alan R. Hildebrand. He became interested in a few KT sites that showed thick, jumbled deposits of coarse rock fragments. They were clustered around the Gulf of Mexico. After several trips to the Texas site, Hildebrand realized that the impact had to be to the south. The site also had to be close, as the Gulf of Mexico was an enclosed body of water and would be protected from a tsunami from a distant impact.[47]

In 1990 Hildebrand visited the Beloc site in Haiti. There he found a layer of greenish brown clay that contained iridium, shocked quartz, and small beads of weathered glass. To Hildebrand and his faculty adviser, William V. Boynton, the clay had the look and composition of material blasted out of a crater. They estimated that the impact could not be more than 1,000 kilometers away, which would place it in the Gulf of Mexico or the Caribbean, consistent with the data from the spherules that pointed to an ocean impact. When Hildebrand and Boynton presented their findings at the 1990 Lunar and Planetary Science Conference, they suspected the crater might be off the coast of Colombia, where a semicircular structure had been seismically detected under the sea floor. Others proposed that the crater was off the coast of Cuba. It was a reporter who told them where to look.

His name was Carlos Byars, of the *Houston Chronicle.* He told Hildebrand of a huge ring buried under the northern Yucatan Peninsula. Byars had heard about it in 1981 from a local geophysicist named Glen Penfield, who had discovered it while working on an airborne magnetic survey of the Gulf of Mexico. In 1978 Penfield was a staff scientist working for the Western Geophysical Company, which had been hired by the Mexican national oil company Pemex. His job was to check on the survey's progress. He would review each day's work at his "inspection station" by the pool at the Maria del Carmen Hotel.

As Penfield went through the long paper strip charts, he noticed what at first seemed to be high-frequency noise. When the strip charts were pieced together, they formed an underground arc offshore, with its cusps pointed south. Penfield realized that this was unusual, as the Yucatan's limestone sediments should not show such a magnetic pattern. Checking further, he asked Pemex for a gravity map of the Yucatan made in the 1960s. The map also showed a semicircle, but turned the other way. The two maps formed a bull's-eye 180 kilometers wide under the Yucatan. At its center was the town of Puerto Chilcxulub. Penfield had long been an amateur astronomer, and he believed it was a buried impact crater.

In 1981, a year after the impact theory was published, Penfield and Pemex's field supervisor Antonio Camargo presented a short paper on the crater at a meeting of the Society of Exploration Geophysicists, which Byars attended. The Chilcxulub crater was the largest impact crater on Earth, but no one noted their discovery. Most of the scientists specializing in impact studies were at a workshop in Utah.

Penfield sought proof that the feature was a crater. In 1952 Pemex had drilled core samples in the area. At a depth of about a kilometer, they encountered a thick layer of what was believed to be a volcanic rock called andesite. The Pemex geologists concluded it was a volcanic dome. Penfield knew that analysis of the core samples would show whether the material was from a volcano or had been melted and uplifted following an impact. Unfortunately, he understood the samples had been destroyed in a warehouse fire. Several years later Penfield tried to find scraps of rock at the old drill sites, even digging through a pig sty. Despite the effort, no fragments were found, and Penfield was at a dead end. After learning of Penfield's discovery, Hildebrand joined forces with him. They were able to track down two drill samples, one from outside the ring and another from near its center. Hildebrand and David A. Kring found shocked quartz in both samples.[48]

Other researchers discovered glass in the KT deposits at Beloc in the form of microtektites, impact-melted glass that had been blasted out of the crater and into space before falling back to Earth. The black glass particles were totally free of small crystals and nearly free of water and gas, which ruled out the possibility they were formed by volcanic activity.[49] The chemical composition of the glass showed it was derived from the continental-crust rocks under the Yucatan. Within the black glass were streaks of yellow glass rich in calcium, which had come from the sedimentary rocks deposited on top of the crust rocks. The yellow streaks also indicated the glass had quickly frozen, a further indication of an impact. Volcanic glass remains liquid for a long time and is nearly always homogeneous.

Analysis of the glass also cleared up the confusion over why the spheroids indicated an ocean impact while the shocked quartz pointed to a continental impact. When the continental crust and continental sediment melted, it had formed a feldspar-rich material that mimicked the chemistry of ocean crust.

In February 1991 Walter Alvarez and several colleagues also found a new tsunami deposit. It was near Tampico in Mexico, about 500 miles from the crater site. The sediment had been laid down some 2,000 feet below the surface of the Gulf of Mexico. The first layer consisted of a mixture of deep-water sediment, impact spherules, and chunks of limestone—debris from the impact. Above that was a layer of marine sediments mixed with coastal sand and fossilized wood. Such materials are completely foreign to deep-water deposits and represented debris torn from the coastal forests of Mexico. Above that were thin beds of rippled sand and fine clay, deposited as the tsunami sloshed back and forth in the Gulf of Mexico. Above and below those layers were the ordinary deep-water deposits of slate and forams.

Those who had opposed the impact theory continued to do so after the discovery of the Chilcxulub crater. Charles Officer told the *New York Times,* "I concede them no victory whatsoever." Officer and his allies challenged all the evidence for an impact in the Caribbean and specifically rejected the idea that the Chilcxulub melt material had been formed by an impact. Yet of four groups to look at the glass particles found at Beloc, only Officer's group denied that they were microtektites formed by an impact.

In late 1991 the Chilcxulub core samples were found. They had not been lost in a fire but had been in storage at the Mexican Petroleum Institute, Pemex's research division. Some of the initial samples were rocks cooled from a melt. They were followed by numerous others, and it was possible to compare them with the material found in Haiti and Mexico. The Chilcxulub melt rocks were soon found by two different groups to date from the KT age. More important, chemical studies showed the same isotope in the melt rocks and the KT tektites from Mexico and Haiti. As the tektites come from Chilcxulub and were found exactly at the foram extinction, the crater must have been formed at the same time.

THE DAY THE WORLD ENDED

After nearly two decades of research and controversy, we now have a basic understanding of the events of 65 million years ago. The object was probably a comet nucleus, some 10 kilometers in diameter, traveling about 30 kilometers per second. In the one or two seconds after it struck the ground, a crater some 40 kilometers deep and 150 to 200 kilometers wide opened, and the vaporized rock and comet debris formed a huge fireball. The hot plume rose right through the atmosphere, sending debris on suborbital trajectories to distant parts of the world. This was followed by a second fireball of carbon dioxide gas, released from the limestone rocks. It scattered added debris as it billowed into outer space.

The sudden loss of a huge volume of rock caused Earth to rebound. The 40-kilometer crater lasted only moments. Its walls began to collapse, while the rocks in the center rose to form a central peak, like those in some lunar craters. The central peak then collapsed, resulting in a series of rings. The shock wave generated by the impact traveled through Earth, creating a kilometer-high tsunami in the Gulf of Mexico, battering the coastlines, and disturbing the deep ocean sediment.

Even as the ground shook throughout what is now Mexico and the United States, the sky began to glow. The debris blasted out of the crater was now reentering the atmosphere. In the process it was heated, and the energy turned the sky into a broiler. The heat set forests alight; the resulting firestorms sent clouds of black ash into the upper atmosphere. Much of North and Central America were devastated by the direct effects of the impact. Europe, Africa, and Asia were less immediately affected, but the destruction was only beginning.

The impact and subsequent firestorms had sent huge volumes of rock dust, ash, water vapor, carbon dioxide, sulfur, and nitrogen compounds into the atmosphere. Within days, these had spread around the world, blocking out the Sun. Earth was plunged into cold and darkness for months. The water vapor probably cleared out of the atmosphere quickly, falling as rain and washing out the dust, but the rain was probably contaminated with both sulfuric and nitric acid, produced when the sulfur and nitrogen compounds formed by the impact combined with water. Water vapor was not the only greenhouse gas produced by the impact. The vaporized limestone released carbon dioxide. As the skies cleared, the temperatures began to soar. The carbon dioxide released in the second fireball now trapped the Sun's heat. The higher temperatures may have lasted thousands of years.[50]

Yet amid the destruction, some plants and animals had survived, areas of life in a burned and poisoned world. The species that survived found themselves in a new world. The ecological niches formerly occupied by the dinosaurs were open. The world we know began to take form. What began with a layer of iridium has led to an understanding of our past. It also gave a warning of one possible future.

THE POLITICS OF IMPACT

The implications of the Alvarez paper were obvious, that asteroids represented the potential threat of human extinction. Between June 9 and 14,

1980, the NASA Advisory Council held its New Directions Symposium. Among the subjects looked at was the threat of asteroid impact. The council examined a detection program that would spot all Earth-crossing asteroids down to some minimum size, determine their orbits, and project their positions 10 or 20 years in the future to identify any impact threats. Should one be identified, the council proposed that a deflection mission be launched, using either a nuclear weapon or a mass driver. The latter would take material from the asteroid and eject it like a rocket exhaust.

The council concluded that technology existed for both the detection and deflection efforts. In addition to the possibility of a large impact, the council also noted that even a relatively small asteroid impact could destroy a major city. Of more concern to the council was the possibility that an asteroid impact could be mistaken for a nuclear explosion and cause a war. The council recommended that NASA establish a detection program with one ground-based telescope and consider an expanded effort with several telescopes, ground-based radar, and spacecraft. It also proposed theoretical studies of impact dynamics, planetary effects, and avoidance scenarios. Finally, the council endorsed a rendezvous mission to an Earth-crossing asteroid, to determine its physical and chemical properties.[51]

The results of the council's recommendations were mixed. In July 1981 NASA organized a workshop titled "Collision of Asteroids and Comets with the Earth: Physical and Human Consequences." The proposal for an asteroid rendezvous mission would become reality a decade and a half later as the NEAR mission. The NASA-sponsored detection program was not approved, however, and searches for Earth-crossing asteroids remained small-scale.[52]

During the early 1980s interest began to develop in asteroid defenses. In 1984 Roderick Hyde, a physicist at Lawrence Livermore National Laboratory, wrote a paper titled "Cosmic Bombardment." Focused on small asteroids and comets, it recommended that a small number of interceptor rockets be kept in Earth orbit. Their nuclear warheads would be kept on the ground until needed.

The event that brought the possibility of an asteroid impact to official attention was the near-miss of asteroid 1989 FC on March 22, 1989. The asteroid passed within 400,000 miles of Earth, less than twice the distance to the Moon and the closest approach since Hermes on October 30,

1937. In the aftermath of the near-miss, the American Institute of Astronautics and Aeronautics (AIAA) recommended that studies be made to detect such asteroids and to find means to deflect them should an impact be imminent. The AIAA brought its recommendations to the House Committee on Science and Technology. As a result, the committee inserted the following language into the 1991 NASA authorization bill:

> The chances of the Earth being struck by a large asteroid are extremely small, but since the consequences of such a collision are extremely large, the Committee believes it is only prudent to assess the nature of the threat and prepare to deal with it. We have the technology to detect such asteroids and to prevent their collision with the Earth.
>
> The Committee therefore directs that NASA undertake two workshop studies. The first would define a program for dramatically increasing the detection rate of Earth-orbit-crossing asteroids; this study would address the costs, schedules, technology, and equipment required for precise definition of the orbits of such bodies. The second study would define systems and technologies to alter the orbits of such asteroids or to destroy them if they should pose a danger to life on Earth. The Committee recommends international participation in these studies and suggests that they be conducted within a year of the passage of this legislation.

Also in 1989 Lowell Wood, a scientist with the SDI program, wrote a follow-up paper, "Cosmic Bombardment II." Like the earlier study, it focused on small asteroids, down to 12 to 60 feet in diameter, "which rain down on our fair planet at rates of dozens to hundreds of strikes per century." Thus as the serious studies were beginning, there was a basic split over what objects were seen as the primary danger.

The NASA International Near-Earth-Object Detection Workshop held three meetings in the summer and fall of 1991. It recommended establishment of the Spaceguard Survey Network, named for a detection system in Arthur C. Clarke's novel *Rendezvous with Rama*. The study proposed construction of six automated telescopes equipped with charged coupled devices (CCDs), in both the northern and southern hemispheres, to scan the skies. Over a period of 25 years, it was estimated that 90 percent of the asteroids larger than a kilometer would be detected.

Initial cost for the six telescopes and the operations center was $50 million. If work was started in 1992, the network would be in operation by 1997. Operating costs would be $10 million per year. It was assumed that Spaceguard would continue to operate once the initial survey was complete. Such long-term activity would be needed to assess the risk from smaller objects and to watch for potentially dangerous long-period comets.[53]

The second of the workshops, on technology to deflect asteroids, was held in mid-January 1992 at Los Alamos National Laboratory in New Mexico. Unlike the Spaceguard workshop, it quickly generated controversy. Editorials attacked the integrity of those supporting an asteroid defense and denied the threat was real. In a *New York Times* op-ed piece, Robert L. Park, a professor at the University of Maryland, referred to the workshop as "a revival meeting for Strategic Defense Initiative true believers." He claimed that the end of the Cold War had "driven the weapons scientists to concoct a new justification for their work." He concluded, "The Star Warriors proposed to defend Earth at stupendous cost against an imagined menace that, if it exists at all, might not threaten Earth for millenniums—or thousands of millenniums."[54]

In the *Washington Post* "Why Things Are" column, Joel Achenbach wrote, "The timing is awfully suspicious—the Soviet Union becomes defunct and almost within minutes there is talk about Earth-crossing asteroids." He added, "What's absurd about this is that even as we worry about hypothetical asteroid collisions we are already doing our best to wreck the planet."[55]

Even more outspoken was an article in *West* magazine, published by the *San Jose Mercury News*. Written by Fran Smith, it was titled "Star Dreck" and began with the question, "Are we in imminent danger of extinction by a doomsday rock? Or is the real danger from another billion-dollar Pentagon boondoggle?" It painted a picture of Edward Teller and other weapons scientists exploiting fear of asteroid impacts to further their agenda of building bigger bombs and exotic space weapons. In one of the article's concluding paragraphs, Smith railed against "any study of doomsday rocks" as a waste of government funds, then wrote, "How can federal laboratories consider bombs to prevent a once-in-a-million-year catastrophe, when the ozone layer is eroding twice as fast as anyone predicted—largely because of chemicals used by the U.S. military? By what twisted priorities would a nation lead a charge against space invaders, while bucking world pressure to combat global warming?"[56]

Behind the editorial outrage was a complex story of the politics of impact. The underlying reason for the controversy was the difference in the cultures of astronomers and weapons scientists. The astronomers saw it in terms of classical research, using the standard ground-based telescopes in the traditional way. Such telescopes were best suited to spotting asteroids a kilometer or larger in size, which would cause extinctions if they hit. The search strategy was also organized in a traditional manner. The telescopes would use existing technology, the funding and goals would be limited, and the time scale would be long-term. David Morrison, who chaired the Spaceguard study, said, "Our philosophy is, you take a complete census [of the kilometer-sized objects], track their orbits, and calculate whether they are a danger. If you find one that is, then you have decades or even centuries to plan a response. You don't need to rush out and build missiles and arsenals." Smaller objects were harder to detect, but these would cause only local damage, so their danger was downplayed. The same was true of the undetected 10 percent of large objects, as well as comets that were not predictable. Morrison said, "My approach is risk reduction, not risk elimination."

That was not how the weapons scientists worked. They had always used cutting-edge technology, on short time scales, and with large budgets. Their interest in the smaller asteroids influenced the scale and type of defense. There would be no time to develop a defense, so it had to be ready beforehand, standing alert. Their focus on the smaller objects also reflected a difference in detection strategy. The weapons scientists wanted to use the new and advanced technology developed by SDI. They also had access to data from the Cold War surveillance systems designed to detect Soviet missile launches and nuclear tests, which indicated that the small objects were the real threat. Because the systems were classified, they could not make public several decades of information; John Rather, who chaired the Los Alamos workshop, could only allude to its existence. He said that because of the classified systems, the Defense Department may "have an idea of the frequency of these things that's even better than the astronomers have."[57]

Besides those differing perspectives, the issue has personal and political dimensions. The astronomers saw the weapons scientists as interlopers who knew nothing about asteroids and held them in contempt. They were called technowizards, obsessed with the "weird and wild" but unable to do arithmetic. The astronomers had spent the previous decade opposing SDI; they loathed Teller and his bombs and distrusted new technology. The weapons scientists returned the astronomers' contempt, regarding NASA as a dinosaur while seeing the future of planetary exploration in the satellites and sensor technology they had developed for SDI.

In the end, there was a problem more fundamental than scientific squabbling and newspaper posturing. Shoemaker called it "the giggle factor." Never in recorded human history had a large impact occurred. No one had ever seen an impact release energy equivalent to dozens of times the power of the entire Cold War nuclear arsenal. No one had seen a fireball so huge it billowed out into space. No one had seen the resulting cloud of debris cover an area as large as Earth. It was easy to dismiss the whole subject as a conspiracy of the Military-Industrial Complex.

The final Spaceguard and Los Alamos workshop reports were formally presented to the House space subcommittee on March 24, 1993. They were both, in the political language of the time, dead on arrival. The controversy had made the subject of asteroid impact a political liability. Col. Pete Worden, deputy for technology at SDI, and John Darrah, chief scientist for the Air Force Space Command, were scheduled to appear but had withdrawn. A source was quoted as saying senior Pentagon officials "didn't want to see headlines that the Air Force was chasing space rocks. The subject looked like it had the potential for a high giggle factor when they are involved in so many larger issues."[58]

No one knew on March 24, 1993, that the whole issue of asteroid impact was about to change. Across the country, at the 18-inch telescope dome at Palomar Observatory, two photos from a series taken the previous evening awaited analysis. Both photos showed the glare from the planet Jupiter, and lost amid the myriad star images was a dark fuzzy line. A faint wispy tail extended to the northwest. It would be another day before human eyes would first see this strange object.

No one had ever seen an impact release energy equivalent to dozens of times the power of the entire Cold War nuclear arsenal. No one had seen a fireball so huge it billowed out into space. No one had seen the resulting cloud of debris cover an area as large as Earth.

Yet.

10

SHOEMAKER–LEVY 9

But the end, when it came, was to be from the sky.
Irresistible. Unimaginable. Mushroom shaped.

The World at War, Episode 22, "Japan"

IN 1969, WITH THE APOLLO landing accomplished, Eugene Shoemaker felt it was time for a new direction in his life. As before, it would relate to impact. In 1962 he had given his astrogeology class at Caltech a homework problem: calculate the number of near-Earth asteroids. At that time only 9 such asteroids had been discovered in the previous three decades. The result was startling. Based on the cratering rates seen on the Moon, there were about 2,000 asteroids two kilometers or larger.[1]

It was clear that a huge number of near-Earth asteroids were undiscovered, and any one of them might hit Earth. Shoemaker decided to undertake a search program. In the fall of 1969 he hired a Caltech astronomer named Eleanor "Glo" Helin to track down all the information about these objects. In 1972 Shoemaker and Helin put together a formal proposal to make the first systematic search for near-Earth asteroids. They planned to use the 18-inch telescope on Palomar Mountain. A Schmidt telescope (named after Bernhard Schmidt, who had originated the concept), it could photograph 8.75 degrees of the sky in a single exposure.[2]

The 18-inch telescope had a distinguished and colorful history. It was the first telescope at Palomar Observatory. Its designer, Russell Porter, was a former arctic explorer who had done much to encourage amateur telescope-making. As a result, Porter was invited to join the design team for the 200-inch telescope, and he originated the split-ring mounting used on the 200-inch. The 18-inch telescope, with its wide field of view, was the perfect complement to the larger telescope. The 18-inch was operational in the fall of 1936. Its small dome was located a few hundred yards from the still-uncompleted 200-inch telescope.

The new telescope was first used by Fritz Zwicky, the legendary Caltech astrophysicist. Zwicky originated the concepts of neutron stars, supernovas, dark matter, gravitational lensing, and clusters of galaxies. As director of research for Aerojet General, an early rocket contractor, he was involved with the early V-2 flights at White Sands. Zwicky even proposed giving the Moon an atmosphere and mining asteroids. He was also arrogant, quick to claim credit, reluctant to share it, and generally impossible to work with. He would assign impossible problems to his students and referred to the other Caltech faculty as "spherical bastards" (i.e., bastards from every angle).

Zwicky searched for supernovas in other galaxies with the 18-inch telescope. His first photos showed supernovas, and the final total was nearly 100. Like Zwicky, the 18-inch telescope could be both brilliant and difficult. Zwicky, who could do one-arm push-ups, often had to push and tug it into position, and its tube was dented with scars of these battles of will. By the time Shoemaker and Helin made their proposal, the 18-inch telescope was little used.[3]

THE PALOMAR PLANET-CROSSING ASTEROID SURVEY

The proposal was approved, and the asteroid search started in early 1973. The original plan was to photograph 250 fields per year. The procedure was to make two exposures of each field, a 20-minute exposure and a 10-minute exposure. The longer exposure was scanned for any asteroid trails; the shorter one served as a comparison. They estimated that two new Apollo and two new Amor asteroids would be found per year. The early days were filled with setbacks; the 18-inch telescope was proving as difficult for Shoemaker and Helin as it had been for Zwicky three decades before. Getting the telescope to work was not the only problem, and it did not seem the search would last out the year.

After six months, Helin found her first asteroid, 1973 NA, which proved to be in an Apollo-type Earth-crossing orbit. It was another two and a half years before the second discovery was made. By 1979 the Palomar Planet-Crossing Asteroid Survey had found 12 new near-Earth asteroids. The low discovery rate was disappointing, and Shoemaker reconsidered his original estimate of 2,000. That estimate could still be accurate if it included asteroids down to one kilometer, rather than the original two kilometers. The 18-inch telescope was hard-put to find objects as small as one kilometer, so Shoemaker sought ways to make it more effective. Shoemaker and Helin had briefly used the larger 48-inch Schmidt telescope on Palomar, but Shoemaker considered the 18-inch telescope more versatile, as it could cover a larger area of sky per observing run.

Shoemaker drew on his geological experience with stereographic images to solve the problem. Rather than taking long exposures and looking for trails, he would take a pair of short exposures about 45 minutes apart, then examine them under a stereomicroscope. An asteroid moving westward would appear as two dots floating above the background stars. If it was traveling eastward, it would appear below the star field. Helin found a company called McBain Instruments to build a stereomicroscope that could hold a pair of 6-inch-diameter negatives. The device cost $10,000, but it soon proved its worth. The first test runs were made in May and June 1980 with encouraging results. The stereo photos could detect asteroids as small as one kilometer.

Another addition to the survey was Eugene Shoemaker's wife, Carolyn. After marrying Eugene in 1951, she became a housewife and raised their three children. With the children on their own, she was restless and joined the Palomar Planet-Crossing Asteroid Survey in 1980. The survey continued for another two years, but in the fall of 1982, Eugene and Carolyn Shoemaker established their own search program. Its goal was to discover enough of the near-Earth asteroids to allow an estimate of their number, which in turn would allow a more accurate calculation of cratering rates. The Shoemaker and Helin groups split time on the 18-inch telescope.

THE SEARCH CONTINUED

After reexamining the procedures used for the previous decade, the Shoemakers decided to make short exposures, eliminate the filters they had been using, and maximize the area of sky covered. The result was an observation assembly line. Each exposure would last 6 to 10 minutes, and only about 2 minutes was required to close the telescope lens cover, remove the exposed film, load a new film holder into the telescope, focus it, turn the telescope to the new field, find the guide star, open the lens cover, and begin the next exposure. The effort required close teamwork between Eugene, Carolyn, and a helper. Each night would produce more than 27 stereo pairs, for a total of nearly 300 exposures during a run of five and a half nights.

Carolyn Shoemaker examined each of the pairs. She quickly proved adept at using the stereomicroscope and picking out real objects from the photographic artifacts. Carolyn found her first asteroid on September 13, 1982. Named 3199 Nefertiti, it was an Amor-type asteroid, with an orbit that crossed that of Mars. She discovered a second Amor asteroid, 1983 RB, the following year; it was recovered a decade later. In September 1983 she found her first comet, Shoemaker 1983p, for the start of a remarkable string.[4] By 1987 she had discovered eight comets, surpassing Caroline Herschel as the most successful woman comet hunter. After another two years she had raised her total to fifteen comets, surpassing William Bradfield as the most successful living comet hunter.

That same year, a new team member joined the Shoemakers. He was David H. Levy, a successful comet hunter and author. Levy's interest in astronomy began when he fell off his bike on the way to a sixth-grade picnic. While recovering from a broken arm, he received a book on the solar system. Inspired by comet Ikeya-Seki in October 1965, Levy decided to hunt for comets. Beginning on December 17, 1965, he searched nineteen years before finally discovering a comet on November 13, 1984. By the summer of 1989 he had discovered five comets and joined the Shoemakers' team. Levy's education was not in astronomy; indeed, he had never taken an astronomy course. Rather he had a degree in English literature from Acadia University in Nova Scotia.[5]

For the next several years the Shoemakers and Levy followed a routine. For seven days each month through the fall, winter, and spring they would observe with the 18-inch telescope. The Shoemakers would drive the 500 miles from Flagstaff, Arizona, to Palomar Mountain. Back at home, Carolyn would examine the photos from the run. After looking at thousands of stereo pairs, she could complete a set in about 20 minutes. The scheduling of the 18-inch sometimes left as little as two weeks between the end of one run and the start of another, meaning that she would finish the photos from one run only a day or two before the next trip to Palomar. During the summer months the Shoemakers would travel to Australia to search for ancient meteor craters.

Carolyn Shoemaker continued to discover comets. In February 1991 her total reached 22 when she found comet Shoemaker-Levy 4, passing the number discovered by William R. Brooks. A year later she surpassed the greatest comet discoverer of all time, Jean-Louis Pons. The one-time doorman at the Marseilles Observatory began comet hunting when he was 40 years old and achieved a final total of 26 named comets. By 1993 Carolyn had reached a total of 30 comets.

They were reminded of the potential for asteroid or comet collisions with Earth during 1989 and 1990, when several asteroids made close approaches. Then in September 1992 comet Swift-Tuttle was observed for the first time in 130 years. When the initial orbit was calculated, it indicated that the comet would approach close to Earth the next time around in 2126. In fact, should the comet make its closest approach to the Sun during a specific 3.5-minute period on July 26, it would then strike Earth on August 14, 2126. Should such an impact occur, comet Swift-Tuttle's orbital velocity (60 kilometers per second) and diameter (5 kilometers) would produce the energy of 20 million megatons of TNT.[6] Further observations refined the orbit calculations, eliminating the possibility of a collision in 2126. Coming only a few months after the controversy over the Los Alamos workshop on asteroid deflection, the editorial response was predictable. *Nature* declared, in an editorial titled "Earth Saved from Disaster," that "Swift-Tuttle is resolutely refusing to be the agent of wild destruction. Too bad!" The November 3, 1992, *New York Times* suggested that impact prediction "is a result of scheming by astronomers and bomb makers by practicing the kind of threat inflation the Pentagon excelled at in the Cold War."[7]

A STRING OF PEARLS

The Shoemaker-Levy team faced a difficult winter at the start of 1993. During the January run there had been only one clear night, and the February run had a total of one clear hour during the seven nights. The run beginning on the night of March 22 got off to a good start. Low clouds blocked out the city lights of San Diego and Los Angeles, and the stars were steady and brilliant. On hand for the run was a French astronomer named Philippe Bendjoya, who was beginning an asteroid observation program and was there to gain experience. The night seemed lost when Eugene developed the first pair of photos; they were black. During the previous month, someone had opened the box of specially prepared film, exposing it to light. It would take six hours to prepare more film. In desperation, Eugene developed some of the film from the bottom of the box. It was light-struck around the edges but usable, and the night was salvaged.

The next night began well, with a fresh supply of film, but high, thin cirrus clouds from an approaching storm moved in after 20 exposures had been taken. Standing outside the dome, Levy thought the clouds were not too thick and suggested continuing. Eugene was worried about the cost, at $4 a shot. Then Levy remembered the exposed film and suggested using the dozen that were left. Eugene agreed, and they returned to the telescope. As Levy looked through the guide scope he had trouble seeing the guide star. The glow from Jupiter nearly washed it out. Two more exposures were taken, then the cirrus moved in. After an hour and a half, a break in the clouds appeared, and Levy rephotographed the Jupiter field and finished the set. Other than several fields the next night, the rest of the March run was clouded out.

The afternoon of March 25 was windy with falling snow. The Shoemakers, Levy, and Bendjoya had assembled at the 18-inch telescope dome. By 4:00 p.m. Carolyn had finished the film from the first night and the few good exposures made on the second. There were no objects of interest in them. She then put the two Jupiter shots under the stereomicroscope. It had been nearly a year since she had found a comet, and as she looked at the negatives, she said somewhat dejectedly, "You know, I used to be a person who found comets." While Carolyn sat at the stereomicroscope, Levy took out his laptop computer and continued work on a book chapter, and Eugene planned the night's observations. Bendjoya was outside, looking for any signs of clearing.

Suddenly, Carolyn straightened up in her chair and said, "I don't know what this is. It looks like . . . like a squashed comet." Eugene studied the images for about a minute, then turned toward Levy, who was sitting by the door. Levy said later that in all the years he had known Eugene Shoemaker, he had never before seen such a look of pure bewilderment on his face. Levy took a look, and then Bendjoya, who had by this time returned. The object did look like a squashed comet. Rather than a single coma and tail, there was a bar of coma, with several tails extending to the north and a thin line extending from either end of the object.

The four of them were not sure what it was, but it did resemble a comet. The find was reported to Brian Marsden at the Central Bureau for Astronomical Telegrams, the official body that announces all astronomical discoveries. Marsden asked if there was someone who could confirm the object. Levy remembered that Jim Scotti was observing that night with the Spacewatch telescope. It was possible that the storm now covering Palomar had not yet reached Kitt Peak. Levy notified Scotti of the object and asked him to look for it. Scotti agreed, even though he thought it was only a reflection of Jupiter. The Shoemakers and Levy were sure the object was real. They spent the next two hours measuring its position, then returned to the 18-inch dome.

They decided to call Scotti. Levy dialed the Spacewatch dome, and Scotti answered. Levy could hardly recognize his voice; he asked, "Are you okay?" Scotti replied, "Oh, yes. That sound you just heard was me trying to lift my jaw off the floor." Levy asked, "Do we have a comet?" Scotti responded, "Boy, do you ever have a comet!" Scotti said there were at least five nuclei visible, and he suspected there were more that could not be seen due to clouds. The nuclei were surrounded by comet material, while the wings on either side were as long or longer than the nuclei train itself.[8]

The announcement of the comet, later designated Shoemaker-Levy 9 (S-L 9), was made the following day. Other astronomers made observations and speculated about the object. At Mauna Kea, Hawaii, Jane Luu and David Jewitt imaged the comet on March 27 with a high-resolution CCD camera attached to an 88-inch telescope. They found seventeen separate nuclei "strung out like pearls on a string."[9] Eugene Shoemaker and Bendjoya thought the comet's peculiar form could have occurred when it passed within Jupiter's Roche limit. The nineteenth-century French mathematician Edouard Roche calculated that if an object held together only by its own gravity passed too close to a planet, the tidal effects would break it apart. A comet could easily be disrupted. Each fragment would drift apart, surrounded by a cloud of debris.

A few days later Jay Mellosh, an astronomer at the University of Arizona, was reading the morning paper when he saw a photograph of the comet fragments. He recalled photos taken by the Voyager spacecraft in 1979 that showed chains of craters on two of the moons of Jupiter. The cause of these crater chains was unknown, as they were not secondary impact craters blasted out by debris from a larger impact. Mellosh concluded they had been created by the impacts of similarly disrupted comets in the distant past.

There was a counterpoint, however, to this saga of discovery. On March 31, 1993, a mere six days after Shoemaker-Levy 9's discovery, the San Diego Public Services and Safety Committee voted to approve the conversion of the city's streetlights to high-pressure sodium. Soon after, one irate letter writer announced, "Little that is done in astronomy anywhere is of any practical importance."

Over the weeks following its discovery, additional observations allowed Shoemaker-Levy 9's orbit to be calculated. By April 3 the picture was clear: the comet fragments were in orbit around Jupiter. They were not in a nearly circular orbit, such as that of Earth's Moon. Rather the fragments were in a highly elliptical orbit that extended from well within the Roche limit out to a distance as far from Jupiter as Mercury is from the Sun. At some point in the past the original comet had been drawn into an unstable orbit around Jupiter. Then, in July 1992, the comet had passed within Jupiter's Roche limit, breaking the nucleus into fragments. The fragments were spotted as they headed away from Jupiter.

That initial orbit was further refined, and by mid-May some of the orbits being calculated indicated that the Shoemaker-Levy 9 fragments might collide with Jupiter. Marsden e-mailed Levy the news, but he and the Shoemakers were about to leave for Palomar. Levy quickly forgot about it.

CIRCULAR NO. 5800

The harsh winter had finally ended on Palomar Mountain, and the night of May 21–22 had proved highly successful. The Shoemakers and Levy had observed until dawn, and so it was well after noon on May 22 when they awoke. After breakfast they went to the 18-inch dome. Carolyn began scanning the previous night's photos, while Eugene was in the darkroom preparing film for the coming night's observations. Levy logged on to check his e-mail and called out, "There are two new circulars about S-L 9." On the screen appeared the text of circular no. 5800, signed by Marsden. In the text, "Delta J" refers to the comet's distance from the center of Jupiter, "AU" refers to an astronomical unit (the distance from the Sun to Earth), and "UT" is universal time. It read in part:

> At the end of April, computations by both Nakano and the undersigned were beginning to indicate that the presumed encounter with Jupiter . . . had occurred during the first half of July 1992, and that there will be another close encounter with Jupiter around the end of July 1994. . . .
>
> This particular computation indicates that the comet's minimum distance Delta J from the center of Jupiter was 0.0008 AU (i.e., within the Roche limit) on 1992 July 8.8 UT and that Delta J will be only 0.0003 AU (Jupiter's radius being 0.0005 AU) on 1994 July 25.4.

Levy said, "Carolyn, our comet is gonna hit Jupiter," and began reading the circular aloud.

From inside the darkroom, Eugene called out, "What! What'd he say?" Outside they could hear lids slamming shut and drawers closing as Eugene put the film away. He opened the darkroom door, ran over to Levy, and began reading the text on the computer screen. All of Eugene Shoemakers' life, his dream of going to the Moon, his work at Meteor Crater, his participation in the Apollo and unmanned lunar and planetary programs, his search for additional impact craters and the two decades of work with the 18-inch telescope, had come together in this moment. Finally, he said very quietly, "I never thought I'd live to see this. We're going to see an impact."[10]

BIG BANG OR BIG FIZZLE?

In the months following its discovery, comet Shoemaker-Levy 9 continued traveling away from Jupiter on its final orbit. Jupiter's gravity slowed it until S-L 9 reached its farthest point on July 16, 1993. The comet fragments were now 50 million kilometers away from Jupiter and traveling at their slowest speed. The gravity of Jupiter held the comet fragments, and they began the long fall back toward the planet—and their destruction a year hence. As they headed back, Jupiter's gravity caused them to speed up. When they reached the planet, they would be traveling 60 kilometers per second, equivalent to 134,000 miles per hour, or Mach 100 (100 times the speed of sound). From the orbital data, it was known that the fragments would enter Jupiter's atmosphere at a 45-degree angle on the night side of Jupiter. As planning began in the summer of 1993, the question became, "Then what?"

The answer depended on assumptions about the size of the fragments and the dynamics of their entry into Jupiter's gaseous atmosphere. Although the speed of entry was known and the density of the comet's material could be assumed to be equal to that of water ice, the size and thus total mass of the fragments could only be guessed. Estimates of the size of the original comet nucleus ranged from 1 or 2 kilometers up to 10 kilometers. An increase of 10 in size (i.e., from 1 to 10 kilometers) increases the total kinetic energy by a factor of 1,000. The levels of energy to be released in the impacts were far beyond any seen in nuclear explosions. A 1-kilometer nucleus of water ice would release energy equal to 250,000 megatons of TNT. A 10-kilometer nucleus, in contrast, would release nearly a billion megatons. A second factor was how this energy would be released when each nucleus hit Jupiter.[11]

The results were three mutually exclusive possibilities. Zdenek Sekanina, Paul W. Chodas, and Donald K. Yeomans, dubbed the quicksters, proposed that the fragments would immediately explode as they hit the atmosphere. A second concept, by Mordecai-Mark Mac Low and Kevin Zahnle (called the pancakers), suggested each fragment would flatten out and penetrate the cloud deck on Jupiter before exploding. More than 90 percent of the fragment's kinetic energy would be released within a fraction of a second. In both cases, a white-hot fireball would erupt above the cloud deck and into space. Upon reaching space, the comet and cloud debris would cool rapidly, become transparent, and collapse back into the atmosphere. It was hoped that the infrared energy would still be detectable a few minutes later when the impact site rotated into view from Earth. There was also the possibility that the cooling debris could condense into large white spots in the atmosphere.

The third possibility, a soft catch, was put forward by Thomas J. Ahrens, Toshiko Takata, and others at Caltech and the University of Arizona. They believed the fragments would remain tightly confined within the surrounding shock front. The debris's kinetic energy would bleed off gradually over several seconds, trapping it within Jupiter's atmosphere. Alexander J. Dessler, a radio astronomer at the University of Arizona, colorfully predicted, "My sense is that Jupiter will swallow these comets up without so much as a burp."[12]

The consensus among most astronomers was that the impacts would cause only minimal effects on Jupiter, for several reasons. It was recognized by all concerned that the unknowns in such an unusual event were so great that reliable predictions were impossible. Under such conditions, it was better to be safe than sorry. That attitude was reinforced by recent experiences with comets. Twenty years before, comet Kohoutek was predicted to be as bright as the full moon, but the early estimates proved to be in error, and comet Kohoutek reached only naked-eye brightness. In 1986 Halley's comet was dim and poorly placed from the United States. The two disappointments made scientists cautious in their estimates.[13]

Whatever the outcome, S-L 9's impact was expected to provide insights into the nature of comets, of Jupiter's atmosphere, and impact dynamics. With the advantage of fourteen months to prepare, a massive observation effort was planned. Mauna Kea, Palomar, and Kitt Peak all rearranged their observation schedules for the week of the impacts. (Time on large telescopes, such as the 200-inch Hale telescope, was scheduled a year and more in advance and was rarely given to planetary observations.) The Kuiper Airborne Observatory, a 36-inch telescope carried aboard a modified C-141 transport plane, was originally scheduled for an overhaul in July 1994. Because its infrared detectors were ideal for observation of any fireball remnants, the maintenance was pushed back to September.

Plans were also formulated to use the Hubble space telescope. Some 100 orbits would be allotted to S-L 9 observations, 36 of them to observations of the fragments before impact. The Galileo spacecraft, heading toward Jupiter, was in a position to image part of the planet's night side, which was not visible from Earth, and it meant that all the impacts could be seen. Two other spacecraft, *Ulysses* and *Voyager 2,* carried radio receivers that could pick up signals generated by the impact. As for amateur astronomers, the range of the predictions made it unclear what they might see, if anything.

With the new year, the questions remained. With its resolution greatly improved by the successful repair mission in December 1993, Hubble had the best view of the fragments as they drew ever nearer to Jupiter. The first new images from Hubble of S-L 9 were returned in early January 1994. They showed the cloud of dust surrounding the fragments but were unable to resolve the objects themselves. Their size would remain unknown until impact.[14]

The Hubble images also showed some unexpected and disquieting changes. The 21 fragments were originally designated A through W. (The letters "I" and "O" were omitted since they could be confused with numbers.) In the original July 1993 Hubble images, fragment Q seemed to be elongated. The January 1994 images clearly showed it had split into two fragments, Q1 and Q2. Even more dramatic was the behavior of fragment P. In July 1993 one fragment was visible. In January 1994 there were now two, P1 and P2. By March 1994 fragment P2 had split again, into fragments P2a and P2b. The evolution of the different fragments, described as "more difficult than following the most involved television soap opera," could also be reconstructed using the Hubble images. The bits and pieces of P and Q had probably come from a single original object. Fragment B was formerly part of fragment C, and F had split from G. Two fragments, J and M, had disappeared entirely.

This continued fragmentation and the disappearances of J and M had implications for the comet's structure and the impact. A number of astronomers suggested that the original comet was of very low density. It could have been a loosely bound rubble pile composed of house-sized blocks with room-sized voids between them. This structure was very weak, and it was suggested by Sekanina that the individual fragments would continue to break up under the stress of rotation and Jupiter's increasing tidal forces.[15]

A week before the impact Paul Weissman published a paper in *Nature* magazine titled "Comet Shoemaker-Levy 9: The Big Fizzle Is Coming." It summarized the rubble-pile model of comets and predicted that the fragmentation seen by Hubble would continue until no large fragments were left to reach Jupiter. They would be reduced to clouds of rocks, gravel, and dust. Rather than a huge explosion, the result would be a meteor shower, as each bit of debris entered and burned up. There would be nothing to see from Earth. Weissman wrote, "The impacts will be a cosmic fizzle. . . . The Shoemaker-Levy 9 explosions may be closer to about 30 megatons each, but still far less than the 100,000 megaton explosions that some have predicted."

Weissman's opinion that S-L 9 would prove to be a big fizzle had already been seconded by Marsden. He had long believed that there were a number of small, one-kilometer comets orbiting Jupiter at any given time. The fact that S-L 9 had broken up favored, in his view, its being a small object. Marsden estimated that such a kilometer-sized comet collided with Jupiter an average of every fifteen years. He stated, "We never see Jupiter affected by these collisions."[16]

It had been fourteen long months since circular 5800 was published. Press interest had steadily increased since the start of the year. By early July it was exceeded only by the press frenzy surrounding the arrest of O. J. Simpson for murder. The telescopes and space probes were all ready, the astronomers were at their observatories, the press had gathered at the Space Telescope Science Institute to await the first Hubble images, and mystics and the tabloids issued their predictions of doom. For the Shoemakers and Levy, the final days had been an endless series of interviews and public appearances. Levy continued to be haunted by the possibility that S-L 9 would be another Kohoutek.

The calculations of S-L 9's orbit had undergone a final refinement. Fragment A would hit Jupiter at 3:55 p.m. EDT on July 16, and the last, fragment W, would impact at 4:34 a.m. EDT on July 22.[17] The first telescope with a chance to see any effects from fragment A's impact was the 3.5-meter telescope at Calar Alto in Spain. The telescope would be pointing at the edge of Jupiter at the time of impact, in the off chance that the fireball might be visible over the horizon. Few astronomers expected to see anything (Eugene Shoemaker being an exception). Fragment A was one of the smallest pieces, and it was impacting at the most distant point from the day side of Jupiter. If any effect lasted the 20 minutes it would take for the impact site to rotate into view from Earth, the Calar Alto telescope could see it. The entire world was now waiting, but no one had any idea what to expect.

FRAGMENT A

Saturday, July 16, 1994, was the twenty-fifth anniversary of the launch of Apollo 11, the first manned Moon landing. That morning, as Levy prepared for another round of interviews, he thought of T. S. Eliot's poem *The Hollow Men:* "This is the way the world ends. Not with a bang but a whimper." Levy hoped not, but with all the predictions of a fizzle, he would be happy if Galileo saw any kind of a flash and Hubble detected even the smallest spot. Despite the instantaneous communications of the late twentieth century, it would be several hours after fragment A's impact time before they knew anything. The Hubble images would not be radioed to Earth until later that evening, while Galileo's stuck main antenna meant that it would be months before its data could be transmitted.

The Shoemakers and Levy were preparing for the 7:30 p.m. press conference at the Space Telescope Science Institute when the first word came in from Spain, in the form of a news flash. Hal Weaver, leader of the Hubble science team observing the comet fragments, heard a radio in another office saying that the Calar Alto Observatory had observed something connected with fragment A's impact. Weaver found Eugene Shoemaker and whispered the news to him. Eugene responded by saying, "You mean they saw a plume?" The Shoemakers and Levy headed to an office to call Marsden, reporters and camera crews in tow. Marsden confirmed that Calar Alto had seen a plume and that it had been confirmed by astronomers at La Silla, Chile. Marsden was in the process of putting out a circular, but the news had already spread. CNN had carried a report, and e-mail was being posted on the Internet. The old methods of scientific publication were being bypassed with S-L 9. New images and observations were posted within hours of being made.

At the 7:30 p.m. press conference, Eugene described the reports of the two observatories but cautioned the reporters that details were still sketchy and needed to be confirmed. Because of conflicts with the space shuttle mission, the download of the Hubble photos had to be delayed for one orbit. It had been nearly four hours since the fragment A impact, and the added delay was fraying everyone's nerves. As the download began, Eugene Shoemaker described a new computer impact model. He showed plots tracing the fireball's rise and then collapse like a pancake onto the top of Jupiter's atmosphere.[18]

One floor below, Heidi Hammel was sitting in front of a monitor with the rest of the Hubble science team, watching the Hubble images. The first two images, taken three minutes apart, showed nothing, but the third looked odd. There seemed to be a little ball peeking over the rim of Jupiter. At first Hammel thought it might be one of the moons of Jupiter and asked Weaver, "Tell me that isn't a Galilean satellite." Weaver checked an almanac and said, "No, it shouldn't be one of the Galilean satellites." The next two images showed the ball growing in size and spreading out. A third image showed the plume flattening out. What Eugene Shoemaker was describing as a theoretical possibility upstairs, they saw in reality. It was followed by images of Jupiter taken one orbit later. They showed a black dot surrounded by a semicircular arc. The spot was debris that had fallen back to the clouds, and it was half the size of Earth. For a moment the scientists sat with smiles of wonderment on their faces. Having feared the big fizzle, they saw a plume rising into space that was even larger than predicted. Someone said, "Look at that." As the realization sank in, they began to applaud and cheer.[19]

The original plan was to release the Hubble images at a second press conference at 10:00 p.m. Looking at the black spot, one of the scientists said, "We ought to take this picture upstairs to the press conference." Another seconded the idea, saying they should do it "right now." A laser print of the photo was made, and Hammel ran upstairs with it and a half-empty bottle of champagne. The press conference was just ending when Hammel appeared at the door. After being introduced, she held up the laser print, and the world saw the way the world ends. It was not with a whimper. She said, "This is in my dreams, the kind of stuff we saw. We're going to have more of them. It's going to be a great week."[20]

Amateur astronomers were also awaiting the impact of fragment A. One of them was Jeffrey A. Beish of Miami, Florida. At 5:30 p.m. EDT, still in full daylight, he pointed his home-built 16-inch telescope at Jupiter. He noticed an unusual darkness near the limb of the planet. He made a drawing before clouds moved in. It was another two hours before the overcast cleared. When he looked again, at 7:35 p.m., he saw an "inky" dark spot. "It had a sort of pigtail and a circular ring around it," he said later. This came as a surprise, as a light-colored spot was originally predicted. Beish immediately began yelling for people to take a look. Carlos E. Hernandez ran out of his house, took a look, and nearly fell off the observing platform as he shouted, "That's it! That's it!" They called Donald C. Parker, who had a telescope equipped with a CCD camera. Over the next week, Parker took more than 1,000 CCD images of Jupiter.[21]

ARMAGEDDON

Fragment B hit Jupiter at 10:50 p.m. EDT, but only a few large telescopes reported a fireball or black spot. The impact was much less spectacular than that of fragment A. The history of fragment B gave an explanation. It had previously split off fragment C, and the lack of a large plume or spot was consistent with the idea that B was a cloud of debris, rather than a large solid object.

July 17 saw the impact of fragments C through F. Fragment C produced a plume and dark spot comparable to that of A. Fragment D, like B, was debris that had split off a larger object, in this case, fragment E. It produced a short-lived fireball and a modest spot. At 11:17 a.m. EDT, fragment E hit Jupiter. The first of the large fragments, it produced the most impressive spot yet. About half as large as Earth, the spot resembled the shadow of one of Jupiter's moons—a small, round black circle with a winglike halo and two spikes. For the final event of the day, fragment F hit nearly 10 hours later, matching the rotation rate of Jupiter so that its smaller spot was close to that of E. This caused confusion as observers tried to sort out the various black spots.[22] The average explosive yield of a typical fragment was estimated to be around 25,000 megatons of TNT. As powerful as these impacts had been, they were only the prologue.[23]

Fragment G hit on Monday, July 18. It was believed to be the largest of the fragments, and for both ground-based telescopes and the orbiting Hubble, its impact was unlike anything ever witnessed by human eyes. The plume extended more than 2,000 miles into space and had a temperature of 50,000 degrees F. Its infrared energy, as bright as Jupiter itself, was so intense that the detectors on several telescopes were saturated. As the plume collapsed, a ring of hot gas spread 33,000 kilometers across the atmosphere of Jupiter. The black spot produced by fragment G was larger than Earth and much blacker than the A, C, D, and E-F spots. There was also a dark crescent to the south.[24]

Eugene Shoemaker estimated that the energy released by the impact of fragment G was equivalent to 6 million megatons of TNT. Had it hit Earth, fragment G would have blasted a crater 36 miles across. At the height of the Cold War, the U.S. nuclear weapons stockpile had a total yield of 20,000 megatons. Soviet nuclear forces totaled about 60,000 megatons. Fragment G released 75 times the energy of all the nuclear bombs in the world.[25]

Over the next two days fragments H, K, and L impacted Jupiter. All three left impressive spots. The spot produced by fragment H was very black with dark material extending north and south. The spot from fragment K initially rivaled that produced by G. The fragment L spot was very conspicuous and strange-looking, with several dark cores and halos. The G and L spots looked to some amateur astronomers like a pair of eyes, with black eyebrows of debris.[26]

July 20 and 21 would see the heaviest bombardment, with ten impacts. Three sets of objects—Q1 and Q2, R, and, finally, S—would each impact one Jupiter revolution apart. The first impact of the two-day series, however, came as a surprise. Fragments J, M, and P1 had disappeared and were assumed to have broken up. The Keck telescope, now the largest in the world, detected several very small spots produced by the missing fragment M. It was followed by fragment N, which left a small spot, and then P1 and P2, which left no traces. Fragments Q2 and Q1 both produced large spots, although not as pronounced as the giant G, H, K, and L spots. Fragments R and S both landed close to the D-G impact point, forming a complex dark feature in Jupiter's atmosphere. Of the other two objects, T left no trace, and U produced only a small spot.[27]

The week of impacts came to a close on July 22, with fragments V and W. The first left no significant traces; W brought events to a close with a bright plume and a dark spot close to that left by fragment K. The K-W region looked like a sunspot, with three dark nuclei and a penumbra of gray material. It was the D-G-S region that was the most impressive, however. A large bruise in the atmosphere of Jupiter, with a long and dark core, encompassed all three impact sites. Strung between them was a series of other black spots, varying in size, shape, and darkness.[28]

These dark spots were a complete surprise. The brownish material was tiny particles in Jupiter's upper atmosphere. During fragment G's impact, Hubble's faint object spectrograph detected emissions of molecular sulfur, carbon disulfide, magnesium, iron, silicon, and possibly hydrogen sulfide. Other telescopes detected hot ammonia, hydrogen cyanide, methane, and ethylene. During several impacts, both water and carbon monoxide were detected. These materials were believed to be derived from both the comet and Jupiter's atmosphere. The intense heat of the fireballs destroyed the original molecules, reducing them to their original elements. As they cooled, the elements recombined into by-products.

From the Hubble images of the A, E, G, and W plumes came another surprise. Despite the differing sizes of the fragments, all four plumes reached the same height of 3,000 kilometers. It was believed that this was due to a balance between the mass of each fragment, the explosive energy it released, and the depth at which this occurred. A more massive fragment released more energy than a smaller fragment, but it also went deeper, so the fireball had to rise through a greater thickness of Jupiter's atmosphere, in effect canceling out the increased energy, resulting in plumes of the same height from fragments of different sizes.[29]

Beyond the scientific results from the S-L 9 impacts, the black spots also made a great emotional impression. On Sunday, July 17, Clark Chapman, a member of the Galileo imaging team, was observing Jupiter from his front yard with his wife and daughter. Chapman was using his childhood 10-inch telescope, which had been in storage for nearly fifteen years. Looking at Jupiter, he was stunned to see the A and E spots. He could not believe how obvious they were. The following night, Monday, July 18, Chapman looked again and saw a spot so big and black that he could not believe what he was seeing. Waves of awe swept over him as he saw the black spot left by fragment G.[30]

That same night, Robert Burnham, the editor of *Astronomy* magazine, and a friend looked at Jupiter with a 6-inch telescope. Looking through the eyepiece, Burnham saw the black spot left by fragment C, "black as night and bigger than I expected." He later described the experience: "The hair on the back of my neck prickled. Why? I think it's because we backyard astronomers get used to seeing celestial objects that always look the same or change only minutely. So when something this drastic happens, it's downright eerie."[31] Even in small telescopes the black spots were prominent. Observing from a Manhattan apartment under hazy skies, George Hripcsak was amazed to see the spots in a 3-inch telescope. Gus Johnson in Swanton, Maryland, reported that he could see the "damaged" areas with a 2-inch telescope.[32]

Over the weeks and months that followed, the black spots were spread out by Jupiter's upper atmospheric winds. They remained the most conspicuous feature on the planet, however. During the first half of 1995 they continued to spread out and grow fainter. Yet even a year later traces of the band could still be seen by some amateur astronomers.[33]

The Shoemaker-Levy team was undergoing changes in the wake of the S-L 9 collision. They continued to use the 18-inch Schmidt telescope until December 1994. The final observing run was actually made in May 1996. It was sponsored by the National Geographic Society, as part of a television special called "Asteroids: Deadly Impact."[34]

The original goal of their search was to calculate the rate of impact. In April 1997 Eugene Shoemaker published the results. Based on the numbers of asteroids and comets found by his and the other surveys, the paper concluded that Earth's cratering rate during the previous 200 million years was possibly twice the rate of the previous 3 billion years. This estimate predicted a 10-kilometer crater every 100,000 years and a 20-kilometer crater every 400,000 years. The rate has accelerated as the solar system's oscillations above and below the galactic plane have diminished, increasing the tidal effects from nearby stars on comets orbiting the outer edge of the solar system, the Oort Cloud. This has caused an increased number of comets to head toward the Sun.

The estimate had a larger error margin than Shoemaker had hoped for, however. Because of the effects of erosion on Earth, craters vanish relatively quickly. He believed that the margin of error was about 25 percent. If, as some astronomers believed, it was closer to 50 percent, there would be no change in the cratering rate.

On July 18, 1997, Eugene and Carolyn Shoemaker were driving a pickup in the Australian outback. For several years they had traveled to Australia looking for ancient meteor craters. As their truck rounded a blind curve, it was hit head-on by a Land Cruiser. There was no chance to avoid the impact. Eugene was killed in the crash, and Carolyn suffered five fractured ribs, a broken wrist, a dislocated shoulder, and other injuries. Eugene Shoemaker's ashes were scattered on the south rim of Meteor Crater.[35]

A small sample was also placed aboard the *Lunar Prospector* spacecraft. It was successfully launched and placed into orbit around the Moon in January 1998. Among *Lunar Prospector*'s first tasks was to confirm the presence of ice at the Moon's poles. After the spacecraft's mapping mission was completed in the early morning hours of July 31, 1999, its engine was fired a final time. The spacecraft fell out of orbit, crashing into a crater near the Moon's south pole, and Eugene Shoemaker reached the Moon.[36]

WATCHING FOR DOOMSDAY

Between July 16 and 22 every nightly news broadcast began with a story from another planet. Through television, the Internet, or a telescope, all of us saw the unimaginable. We have seen impacts dwarfing Earth's nuclear arsenals, we have seen fireballs climbing into space, we have seen debris spread over areas larger than Earth, and we have watched dark spots linger for a year. What happened to Jupiter could also happen to Earth. If it should, human existence as we know it would end. In the wake of S-L 9, the question of keeping a systematic watch for near-Earth asteroids was reexamined.

One early realization was that improvements in CCD technology would allow smaller, existing telescopes to meet the goals originally set in the Spaceguard study. After S-L 9, the organizational framework needed to support a search effort was set up. On March 26, 1996, the Spaceguard Foundation was established in Rome, Italy, to encourage international participation. Complete sky coverage required that observatories in both the northern and southern hemispheres participate and that they be spread out over a wide area. The latter would prevent several sites' being covered by the same weather front. Coverage in Japan was considered particularly important.[37]

Another realization was that the U.S. military already had a surveillance network adaptable to detection of NEAs, the U.S. Air Force's ground-based electro-optical deep space surveillance (GEODSS) telescopes. The 39-inch telescopes were designed to detect and track satellites in high Earth orbit. A joint effort between the U.S. Air Force, NASA, and JPL was organized. Called Near-Earth Asteroid Tracking (NEAT), it used the GEODSS telescope on Mount Haleakala, Maui, Hawaii. JPL designed and built the NEAT camera and computer system, which was then installed on the Air Force telescope. Eleanor F. Helin was the principal investigator, with David L. Rabinowitz and Steven H. Pravdo as coinvestigators.

NEAT began operations in December 1995. With its more sensitive CCDs, short exposure times, and automated detection software, NEAT could detect fainter objects faster than the 18-inch Schmidt telescope Helin had been using. The observation initially ran twelve nights each month (later reduced to six nights), centered on the new moon. The initial funding from NASA was $1 million per year.

Within months NEAT was proving its potential. Between December 1995 and late April 1996 it detected more than 2,400 asteroids, about 55 percent of which were new objects. Of the new discoveries, more than 200 received provisional designations. The March 1996 observing run was the first "good weather" run and detected more than 1,000 asteroids, including high-inclination inner-belt asteroids and potential Mars-crossers. There were also 4 new NEAs detected, half of all the NEAs discovered during the month of March 1996. Two of the asteroids were considered potentially hazardous asteroids (PHAs), defined as asteroids 110 meters or larger in diameter with orbits such that they could collide with Earth sometime in the future.[38]

In late April 1996 a new very large CCD array (4,096 pixels square) was fitted to the NEAT camera, and by April 1998 NEAT had achieved some remarkable statistics. The total number of asteroids detected was 23,061. The number of new objects was 12,025, just over 52 percent of the total, of which 1,488 received new designations. The number of NEAs was equally impressive, 16 Amor asteroids and 9 Apollo-type Earth-crossers. Of those, 6 were PHAs. One, 1998 HD14, was an Aten asteroid, some 270 meters in diameter, discovered on April 26, 1998. Of the 30 known Aten asteroids, 1998 HD14 was the fifth found by NEAT. Aten asteroids, because they orbit inside the orbit of Earth, were considered the most likely to impact Earth. They are never far from Earth's orbit, and they may cross it as many as four times a year.[39]

At the same time, the long-established Spacewatch project at Kitt Peak was continuing. In contrast to NEAT, it is a full-time effort, observing 20 nights per month with the 0.9-meter Steward telescope. (The observing run is interrupted by the full moon, which makes the night sky too bright.) By March 1998 it had found 160 NEAs, 16,000 new asteroids, and 17 new numbered asteroids. In early 1998 a new 1.8-meter telescope was under construction for the Spacewatch project. It would increase the sky area covered and be able to detect fainter asteroids.

David Tholen, at the University of Hawaii, undertook a unique search effort. Rather than looking in the region of space almost opposite the Sun, he looked at the dawn and dusk skies with the 2.24-meter telescope atop Mauna Kea. His targets were asteroids within Earth's orbit. In February 1998 Tholen discovered 1998 DK36, the first of a whole new class of asteroids. Its orbit ranged from just inside the orbit of Earth to as close to the Sun as the orbit of Mercury. All other asteroids, including the

Atens, have at least part of their orbit outside Earth's. The asteroid is thought to be about 40 meters in diameter, roughly the same size as the Meteor Crater and Tunguska objects.

Tholen noted that 1998 DK36 would not have been found by other search efforts, as it never traveled in the parts of the sky they scan. The closest it approaches Earth is estimated to be 750,000 miles. If it had been on a collision course, however, it would have struck from the daylight sky, without warning. Tholen added, "To do a better job with such discoveries, we really need to have a telescope that we can dedicate to such difficult observations." He continued, "1998 DK36 is nothing to lose sleep over. It's the ones we haven't found yet that are of concern."[40]

Other programs include a search effort at Lowell Observatory, the Catalina Sky Survey by the University of Arizona, and Spaceguard Canada, a joint effort by the University of Victoria and the Dominion Astrophysical Observatory. The Canadian effort uses a 0.25-meter Schmidt, a 0.5-meter, and a 1.8-meter telescope to produce real-time confirmation, recovery, and follow-up of newly discovered NEAs. The Klet and Ondrejov observatories in the Czech Republic are also involved in confirmation and recovery of NEAs. The Beijing Astronomical Observatory made four NEA discoveries between April 1995 and March 1998; two of the four were classified as PHAs. A French asteroid survey is being conducted by Alain Maury from Cote d'Azur.[41]

NEA searches are not only undertaken by professional observatories. The technology available to amateur astronomers has vastly improved since the original Spaceguard study. On June 28, 1997, Roy Tucker discovered the first Aten asteroid found by an amateur. He used a 14-inch Celestron telescope with a homemade CCD camera and observed from his backyard site near Tucson, Arizona. To filter out most of the main-belt asteroids, he observed areas away from the ecliptic. He then compared the pairs of images to detect any moving objects. He searched only 28 hours, taking only 83 image pairs, before finding 1997 MW1. The discovery made Tucker the winner of the first Benson Prize from the American Astronomical Society, an award established by James W. Benson, chairman of the Space Development Corporation, to spark interest in Earth-crossing asteroids. Such amateur activities could substantially increase existing capabilities.[42]

NEA search efforts were also gaining official support. In early 1998 work began on establishment of a new program office at NASA, focusing on asteroid surveys and studies. The program office would integrate all of NASA's programs and research involving NEA. NASA also works with the Air Force Space Command and the National Reconnaissance Office, through the Joint Agencies Partnership Council. The goal is to discover 90 percent of all NEAs a kilometer or larger within ten years.[43]

Of the post S-L 9 asteroid search programs, none has dominated the field like the Lincoln Near-Earth Asteroid Research (LINEAR) project. It uses a 1-meter telescope, on the White Sands Missile Range, close to the Trinity Site, where the first A-bomb test was made. The initial tests were made in early 1996, with additional field trials in January 1997 to qualify the systems. LINEAR went into operation in March 1998 and brought about a revolution in asteroid detection. The change is comparable to the replacement of visual searches by photographic surveys.

Before LINEAR began operations, the Minor Planet Center received a total of about 10,000 position reports of asteroids and comets per month. In a single month LINEAR produced 160,000 or more observations. Between March 1998 and October 1999 LINEAR had produced 1,262,762 observations and detected 273,767 asteroids. This resulted in 26,847 new designations, surpassing the number of asteroids discovered by Spacewatch in its first sixteen years. LINEAR also found 277 confirmed new NEAs.[44]

Those results are achieved through automated operation, a new CCD camera, and search software. LINEAR automatically runs through its search pattern in the sky—all the astronomer has to do is take the cloth cover off the telescope. The advanced CCD camera takes a 10-second exposure, then sends the data to the search computer, and repeats five times for each section of the sky. The computer then compares the images, leaving only the asteroids and comets. The five exposures give a high confidence of detection and a low rate of false alarms. In ten nights of observation per month, LINEAR can cover the whole sky. Grant Stokes, an astrophysicist who oversees the project for the Air Force, commented, "The Minor Planet Center tells us that when we sweep through an area we find everything we're supposed to find: everything they know about, plus stuff they don't know about."[45]

LINEAR was to be one element of an expanded search program. The initial action was an expansion of the LINEAR Project to a pair of telescopes in New Mexico, each of which would operate eighteen nights per month. NEAT operations were also to be expanded from the current six nights to eighteen nights per month. In the longer term, the NEAT detector would be transferred to a 1.2-meter Air Force telescope on Maui, and possibly additional Space Command telescopes would make asteroid searches.[46] By mid-1999 the expansion plans had changed. Rather than an Air Force telescope, the 48-inch Schmidt telescope on Palomar Mountain would be modified with a CCD camera and computer control system. Originally built in 1949, this telescope has several advantages over the NEAT telescope. Because it is larger, 48 inches (1.2 meters) as opposed to 39 inches (1 meter) for NEAT, the Schmidt could detect fainter asteroids. Its field of view is also ten times wider than NEAT's. Initial tests with the NEAT CCD camera attached to the 48-inch Schmidt were made on June 9 and 10, 1999, and proved successful.[47]

But ground-based optical telescopes are not the only source of data on NEAs.

THE SECRET DATA

On March 24, 1993, the day before the discovery of the S-L 9 comet, John Rather appeared before the U.S. House of Representatives subcommittee on space. Rather had been chairman of the Los Alamos Workshop on Near-Earth Object Interception. The workshop had focused on smaller objects, rather than the kilometer-sized asteroids that were the subject of the Spaceguard Survey, a difference that sparked a bitter controversy.

Behind that difference was a conflict in estimates of the number of small asteroids entering the atmosphere. Only a few reports of high-altitude explosions of meteors were known. A brilliant night-time fireball was seen above British Columbia in April 1965, and a daytime meteor had grazed the atmosphere over the Rocky Mountains in August 1972. Yet Eugene Shoemaker had estimated in a 1983 paper that an asteroid should explode in Earth's atmosphere with a yield of 20 kilotons about once a year. A decade later, using the Spacewatch data, David Rabinowitz estimated that a 20-kiloton asteroid air burst should occur every month, and several hundred kiloton-sized bursts should happen per year. Such large explosions should be noticed, but were not—publicly.[48]

In his testimony Rather made an oblique comment. After noting the recommendation to build the Spaceguard telescopes and expressing support for it, Rather said, "Participants of our Interception Workshop who were cognizant of presently classified technical capabilities opined that important immediate progress can result from sharing existing defense search, tracking, and homing technologies with the civilian sector and from implementing protocols to transfer data on NEO [near-Earth object] discoveries from defense and intelligence assets to appropriate centers for determining precise orbits and other relevant data."[49]

With the realization that S-L 9 was going to impact Jupiter, the secret data on NEOs were made public. In October 1993 the Department of Defense declassified data from the Defense Support Program (DSP) satellites, a series of early warning satellites in geosynchronous orbit that carry large infrared telescopes to detect Soviet missile launches. They are also equipped with flash detectors to spot atmospheric nuclear tests. Between 1975 and 1992 the DSP satellites had spotted 136 explosions of asteroids in Earth's atmosphere, with yields ranging from 500 tons to 15 kilotons. Work on the DSP asteroid detection program began in 1975 in order to separate asteroid air bursts from nuclear explosions.

Two of the largest explosions occurred on April 22, 1988, and October 1, 1990. The first fireball, over Indonesia, rivaled the Sun in brightness, but for less than a second. The other, over the western Pacific, was caused by a stony, 100-ton asteroid exploding at about 30 kilometers altitude. It was considered a possible nuclear explosion, and only several months of analysis identified it as an asteroid air burst.[50]

Simon P. Worden, an astronomer who had headed part of the SDI program and who had arranged the release of the data, noted that the October 1, 1990, explosion had occurred during the buildup to the Gulf War. An observer on the ground would have seen a part of the sky become as bright as the Sun, then heard a loud, low rumble. It would have looked just like a nuclear explosion. Worden said, "Had this occurred over Kuwait, it could have been a sticky situation. We could tell it was natural, but they could not."

The DSP data indicated an average of eight asteroid air bursts per year, below the rate predicted by the Spacewatch data on small NEAs. This discrepancy was due to the design of the satellites and the data-handling procedures. A DSP satellite is spun so the telescope scans across Earth. As an asteroid flash lasts only one or two seconds, the telescope will miss four events for every one it sees. Another reason is that the DSP ground controllers look for missile launches and nuclear tests; events outside those categories are missed or ignored.[51]

A few months after the declassification, the brightest asteroid air burst on record was detected, on February 1, 1994, over the Pacific Ocean southeast of Kusaie Island. The fireball reached a magnitude of −25. Two fishermen who witnessed the event reported seeing a very bright reddish and bluish flame. There was no sound, and the meteor's smoke trail lasted an hour.[52] Out in space, six satellites detected the flash. The amount of energy released was a matter of debate. David Morrison and Kevin Zahnle of the NASA Ames Research Center calculated the explosive yield at about 110 kilotons. The original asteroid was estimated to be a stony meteor about 15 meters in diameter, which had exploded at an altitude of about 20 kilometers.[53]

Concurrent with the release of the DSP data was the declassification of acoustic measurements from ground stations, made between 1960 and 1974 by a worldwide network of very low frequency microphones. Operated by the U.S. Air Force, the network was designed to detect shock waves from atmospheric nuclear tests. Some of the microphones were located on the roofs of U.S. embassies. They recorded a few large air bursts. On September 26 and 27, 1962, two separate air bursts occurred over the Middle East. The two asteroids, about 25 and 20 feet in diameter, produced explosions of 30 and 20 kilotons. The largest recorded by the acoustic network occurred on August 3, 1963. An asteroid some 80 feet across entered Earth's atmosphere between Africa and Antarctica, exploding with a yield of 1 megaton. The April 1, 1965, British Columbia fireball was also detected by the acoustic network. The data were sufficient to plot the trajectory, and an attempt was made to find the crater, but the 6-meter asteroid had exploded at high altitude.

The worldwide acoustic system was shut down in 1974, replaced by satellites and seismometers. A decade later, in 1983, Los Alamos began operating its own system of acoustic sensors to monitor underground nuclear tests. The equipment was located in New Mexico, Utah, Nevada, and Wyoming. Under ideal conditions, the four sites could detect acoustic signals from meteors as small as a few centimeters across. The array detected the February 1, 1994, fireball over the Pacific. The acoustic network also detected a swarm of nearly a dozen large fireballs between October 2 and 4, 1996. At least five fireballs were seen over California, two over New Mexico, and several over the Pacific Northwest.

Two fireballs on October 3 attracted the most attention and speculation. The first was seen over New Mexico and Texas. About 105 minutes later another fireball was seen above Little Lake, California, at 8:45 p.m. PDT and was reported from Los Angeles and San Francisco. There was speculation that a single meteor had entered the atmosphere, skipped out into space, made one orbit of Earth, and reentered over California. From the acoustic data, Douglas O. ReVelle at Los Alamos concluded that they were actually two different objects. The entry angle of the New Mexico fireball was such that it could not have skipped back out into space. ReVelle noted, however, "There are a number of questions left to be answered about the October 3 fireballs, and there are some things which don't quite add up. You know, I'm not really sure what was happening in the sky that night."

Additional acoustic data from the Los Alamos network were declassified in December 1996. Based on both the 1960–1974 and 1983–1996 data, ReVelle and his colleagues concluded that each year at least one asteroid explodes in the atmosphere with a yield of 15 kilotons or larger. There are also ten to eleven air bursts per year in the 1-kiloton range, for a total of twelve events per year. That estimate matched Shoemaker's prediction but was still well below the Spacewatch estimate.[54]

The Los Alamos array subsequently detected a bright daylight fireball seen and heard over New Mexico and Texas on October 9, 1997. Witnesses reported it was as bright as the full moon or the setting sun. The sound it produced was likened to a freight train. ReVelle said, "The meteor made a huge sonic signal." It was estimated to be 0.5 to 0.75 meter across and to have released explosive energy equal to about 500 tons of TNT. Data from Los Alamos indicated a possible impact point near El Paso, Texas.[55]

This fireball was also detected by satellites, as it exploded at an altitude of 36 kilometers near El Paso. On December 9, 1997, Air Force satellites detected the breakup of a large meteor over Greenland into at least four pieces. The first exploded at an altitude of about 46 kilometers; the other three exploded close to each other, at altitudes between 28 and 25 kilometers.

Another asteroid air burst was detected over the United States on January 11, 1998, roughly midway between Denver and Grand Junction, Colorado.[56] Such acoustic detections of asteroid air bursts are seen as more complete than the satellite data. The very low frequency sound waves are easier to detect than the flashes of light.[57]

The DSP and acoustic data provide a benchmark to test estimates of the number of small asteroids entering Earth's atmosphere. This, in turn, can be used to estimate the number of small NEAs, which would indicate the effectiveness of the telescope searches. But the current and possible future search programs, both by ground-based telescopes and military systems, are only half of the equation. There remains the question of what to do if an incoming asteroid or comet is discovered.

11

PLANETARY DEFENSE

Clearly, Icarus must be stopped. No effort or funds will be spared in carrying out the detailed plan to be developed by the crack team of scientists and engineers assigned to the project . . . it must succeed.

Paul E. Sandorff, Course 16.74 announcement, MIT, 1967

THE IDEA OF A planetary defense to protect Earth against asteroid or comet impacts has a long history. The father of planetary defense was Lord Byron. The idea of a comet impact was a recurring theme in his writings. He also realized that with technology more advanced than that available in the early industrial revolution, it might be possible to destroy a comet heading toward Earth. While at Pisa in 1822 he suggested, "Who knows whether, when a comet shall approach this globe to destroy it, as it often has been and will be destroyed, men will not tear rocks from their foundations by means of steam, and hurl mountains, as the giants are said to have done, against the flaming mass?—And then we shall have traditions of Titans again, and of wars with Heaven."

The idea also occurred to Allan O. Kelly and Frank Dachille in their 1953 book *Target Earth: The Role of Large Meteors in Earth Science.* After suggesting that an impact had killed the dinosaurs, they noted that humanity might need to develop a means to protect itself against a similar fate. They wrote, "This system will require perpetual surveillance of a critical envelope of space with the charting of all objects that come close to a collision course with the Earth. It will require, further, that on the discovery of a dangerous object moves be made to protect the Earth."[1]

They lacked a detailed analysis of the technological requirements and the physics of defending against the impact of a large asteroid. It was not until the close approach of the asteroid 1566 Icarus in 1968 that this was done. The analysis was not a government study, however, but rather a student project.

PROJECT ICARUS

Shortly before the start of the Massachusetts Institute of Technology spring 1967 term, an unusual course announcement was posted on bulletin boards across the campus. The class was Course 16.74, Advanced Space Systems Engineering, and its students were to develop a plan to save the world. The announcement noted that in June 1968 the asteroid 1566 Icarus would pass within 4 million miles of Earth. In the scale of the solar system, that was a narrow margin. The announcement continued, "The project to be handled by the Advanced Space Systems Engineering students this term assumes that Icarus will, in fact, collide with the Earth."

Twenty-one seniors and graduate students registered for Course 16.74. Their instructor was Paul E. Sandorff. The announcement had been greeted with a certain amusement on campus. Among the suggestions were "How about building a big trampoline?" or "Why not move the Earth out of the way?" To a certain extent, the students shared this attitude. Soon, however, they were caught up in the enormity of the task.

The students were organized into seven groups: orbits and trajectories, nuclear payloads, boosters and propulsion, spacecraft, guidance and control, communications, and economics and management. Because the decisions of any one group affected all the others, there was an absolute requirement for close coordination between them. The project report later said, "This iterative interplay between intragroup and intergroup problem solving was the most significant contribution of the course to our education and experience."[2]

In planning their defense against Icarus, time was the most critical factor. For the purposes of the study, it was assumed that Icarus would strike Earth at 12:26 p.m. on June 19, 1968, in the mid-Atlantic, at a point about 2,000 miles east of Florida. Icarus would be traveling at a speed of 18 miles per second, and the impact of its 4-billion-ton mass would release energy equivalent to 500,000 megatons of TNT. This would blast a crater in the ocean floor 15 miles across, generating tsunamis that would wash away the resort islands, swamp most of Florida, and hit Boston with 200-foot-high waves. The coasts of Africa and Europe would also be hit. In addition, the huge amount of debris and water vapor blasted into the upper atmosphere would disrupt Earth's climate, potentially for decades. Earthquakes a hundred times more powerful than any ever recorded would be felt worldwide. Under the conditions of the study, there were only 70 weeks to prevent this disaster.[3]

Several possible mission profiles were examined by the students, including a soft landing of a rocket on Icarus that would change its orbit so it would miss Earth, implanting a nuclear weapon below the surface of the asteroid to deflect it or break it up, disintegrating it with a high-speed nuclear-armed interceptor, or using a high-speed interceptor to explode a nuclear weapon close to its surface to change the asteroid's orbit.

Ideally, any attempt to change the orbit of an asteroid should be made when it is farthest from the Sun, at its aphelion, where the asteroid's speed is lowest. A small velocity change (called Delta V) would result in a major change in the orbit. The conditions of the study put the aphelion out of reach. To reach Icarus at its November 1967 aphelion, the launch would have to take place in February 1967, i.e., a few weeks after the class started. A flyby mission could be launched as late as October 1967, but even that was ruled out by time constraints. Thus, any defense of Earth would have to be by a nuclear-armed, high-speed interceptor.

The question then became whether to try to destroy Icarus or to deflect its orbit. The best guess was that Icarus was between 2,600 and 5,000 feet in diameter, with the most probable value being 4,200 feet. The mass estimates ranged between 380 million tons and 17 billion tons, depending on whether it was made of comet material or nickel-iron. The most probable mass was 4.4 billion tons, based on the belief that it was similar to a stony meteor. Destroying an asteroid of that size and mass was impractical, as the crater depth had to equal the asteroid's diameter. This would require a nuclear weapon with a yield of 1 billion tons of TNT, far beyond the state of the art of weapon design and the capability of any possible rocket to carry it. Distributing the total yield among several nuclear weapons was ruled out by the timing of the detonations; the first weapon to detonate would destroy the others before they were triggered.

The deflection profile had its own stringent demands, however. The farther from Earth the deflection was made, the less Delta V was required. At a range of 20 million miles, a Delta V of about 20 feet per second was required to cause a total deflection of 4,000 miles (one Earth radius). By the time Icarus was within 5 million miles of Earth, the figure had climbed to 100 feet per second. At a range of 2.5 million miles, the Delta V would be 400 feet per second.

Given the mass of Icarus, a very high yield nuclear weapon was still required to provide the needed Delta V. The study assumed a 100-megaton weapon would be carried by the interceptor. The only booster capable of carrying such a payload was the Saturn V. The students looked at various advanced versions of the Saturn V, including launching two boosters and docking the two S-IV stages in orbit, as well as a Saturn V with a Centaur upper stage. These could provide the maximum payload, but the short development time ruled them out.

In the end, the study determined that a basic Saturn V would have to serve as the booster. Production of the Saturn Vs could be speeded up through added personnel and a three-shift schedule, to deliver nine boosters by the start of interceptor launches in early April 1968. Three of the Saturn Vs would be used for test launches; the remaining six would launch the interceptors. To support the salvo launches, a third pad, Launch Complex 39C, would be built. This would allow launches to be made at intervals of two weeks or less.

THE ICARUS INTERCEPTOR

The interceptor would be built around the Apollo service module, with modifications to carry the 100-megaton warhead. The complete assembly was 42 feet long and weighed 100,000 pounds. As with every other aspect of the project, the modifications were limited by the deadline. As part of the Apollo Applications Program (later renamed Skylab), onboard power systems were then under development to allow the Apollo spacecraft to operate for up to 60 days. That put an upper limit on the flight time.

The service module itself was cylindrical, 12.8 feet in diameter and 24.6 feet long. It contained the fuel and oxidizer tanks for the service propulsion system rocket engine. Three fuel cells provided electrical power for up to 60 days. Should there be a temporary fuel-cell problem early in the mission, or a complete failure during the terminal phase of the intercept, two silver-zinc oxide batteries would provide backup power. Mounted on the outside of the service module were four sets of reaction control thrusters, which would stabilize the spacecraft during the cruise to Icarus, as well as orienting it during course corrections. Communications to and from Earth were via a high-gain antenna at the base of the service module. The temperature of the onboard systems was controlled by radiators mounted on the outside of the vehicle. The service module was not the optimum size for the mission, and it provided more than the required guidance capability. Using it, however, avoided the problems of integrating a completely new payload to the Saturn V booster.

Attached to the forward end of the service module was the payload module. It contained the 100-megaton nuclear warhead. In 1967 unclassified information about the design of very high yield nuclear weapons was limited. The study assumed such a weapon would weigh 40,000 pounds and be a cylinder 140 inches long and 36 inches in diameter. The warhead was mounted horizontally within the payload module. It was to be inside a 4,000-pound casing that would protect it in the event of a launch pad failure or suborbital abort, preventing it from spreading radioactive debris on the pad or in Earth's atmosphere.

The weapon would also require auxiliary systems. It was imperative that the weapon not detonate accidentally. To prevent this, the weapon would be fitted with a safe-arm system. Launched in an unarmed condition, it would not be armed until specific criteria had been met, such as the target acquisition signal and a radio command from Earth. It was also necessary that there be a means to destroy the nuclear weapon should the mission fail, for instance, if an S-IVB failure left the interceptor stranded in Earth orbit or there was a systems failure during the cruise phase. An explosive destruct system would be triggered to blow the nuclear warhead apart, preventing any possibility of a nuclear explosion. The destruct charge would be activated either by a command from Earth or when a timer indicated the planned interception time had been exceeded.

Mounted outside the payload module was the phased-array Icarus tracking radar antenna and Earth navigation antenna. These would be used during the final intercept phase. Radar signals from Earth would be transmitted toward the interceptor and Icarus. One of the interceptor's antennas would pick up the signal directly from Earth, while the other would detect the signals reflected from Icarus, providing precise tracking data for the terminal corrections.

Atop the payload module was the command module, a lightweight shell the same size and shape as the Apollo spacecraft. As the interceptor was unmanned and on a one-way mission, there was no need for the life-support systems and heat shield of the manned spacecraft. At the forward end of the command module was a Sun sensor, and star and Icarus optical trackers were mounted in its side. During launch they were protected by covers, which would then be jettisoned once the interceptor was in space.

Because of the 125,000-feet-per-second closing speed, it would not be possible to use an impact fuze to detonate the nuclear weapon. Rather, it would be triggered by a radar signal when the interceptor reached its closest point to the asteroid. The primary radar was mounted in the side of the command module, with the backup on the forward end.

PROJECT ICARUS ATTACK PROFILE

The six launches would be divided into four high-altitude intercepts and two low-altitude missions. In each case, the attack profile remained the same. After launch the S-IVB third stage, with the attached interceptor, would be placed into a low parking orbit around Earth. At the calculated time, the S-IVB's rocket engine would reignite. After its fuel was exhausted, the engine would shut down and the interceptor would separate. The interceptor's own engine would then be fired to provide the final boost. With the added weight of the nuclear weapon, the S-IVB did not have the ability to send the interceptor on a direct trajectory to Icarus. Some six hours after leaving Earth orbit, a midcourse correction would be made, placing the interceptor into its final trajectory.

Saturn Icarus-1 (SI-1) would be launched April 7, 1968, 72.9 days before impact. SI-2 would follow on April 22 (58.3 days before impact), SI-3 on May 6 (44 days), and SI-4 on May 17 (32.7 days). The date of the first launch was set by two factors, the range of the ground-based radar that was to illuminate Icarus and the maximum operating lifetime of the service module. The other launch dates were based on the time required to check out the rockets and repair the pads following the launches.

On June 6, 1968, SI-1 would reach Icarus, 13.9 days before impact and 60 days after launch. The distance between Icarus and Earth would be 20 million miles. (The flight time and range were both the maximum possible within the limitations of the system.) SI-2 would arrive on June 9, SI-3 on June 12, and SI-4 on June 14 (the latter only 4.9 days before impact). During the first series of intercepts, the range to Earth was closing very rapidly. The first was at 20 million miles, the second at 15.5 million, and the third and fourth 10.8 million and 7.7 million miles, respectively.

The actual intercept was a very demanding problem in accuracy. There were to be three terminal corrections. The first would be made 1.8 hours before interception, when the interceptor would be 150,000 miles from Icarus, and the optical tracker would have reduced the cross-track uncertainty to about 7 miles. The second terminal correction would be made at a range of 5,000 miles, only 3.5 minutes from intercept. At this point, the uncertainty would be reduced to 0.25 mile. The second and third terminal corrections would use data from both the optical tracker and the onboard radar antennas. The final terminal correction would be made at a range of 1,200 miles from Icarus, only 50 seconds before interception. This was to bring the interceptor to within 100 feet of the Sun-lit face of Icarus. Five seconds before intercept, the fuzing radar would acquire Icarus, and the weapon would be armed. When the fuzing radar determined the range was down to 550 feet, the firing command would be triggered. The weapon would then detonate 100 feet or less from the asteroid's surface.

Such precision was required by the particularities of a nuclear explosion in space. Some 70 percent of a nuclear weapon's energy is released in the form of X-rays. When these X-rays struck the surface of Icarus, they would be absorbed and the energy released in the form of heat, blasting a shallow crater, 1,200 feet across but only 15 inches deep, in its surface. The total mass blown out of the crater would amount to 26,455 tons. Icarus would be pushed sideways by the thrust of the expelled mass.

The exact amount of Delta V that would result was not clear, due to uncertainties about the asteroid itself and the lack of unclassified information about the physics of nuclear explosions in space. The most conservative estimate was 26.25 feet per second, sufficient to deflect Icarus at the 20-million-mile maximum range with a single interception. At closer ranges, it would take two or three interceptors. A low-altitude burst was critical to the success of the deflection attempt. The diameter of the crater, and thus the mass ejected from it, dropped sharply as the distance from the surface increased. At a distance of 1,500 feet, the crater would be only 300 feet across. Just as the size of the crater decreased, so too would the resulting Delta V.

During the month before impact, the asteroid's trajectory would be monitored by ground-based telescopes. During the first series of intercepts, their observations would indicate if the deflection attempts had been successful. Once it had been confirmed that Icarus was no longer on a collision course with Earth, the destruct systems of the remaining nuclear warheads would be triggered, removing any chance that they might deflect Icarus back toward Earth.

The attack plan became more complex should Icarus be fragmented. The remaining interceptors would be directed against any fragments still heading toward Earth. Large fragments could be identified within a day after each interception. The smaller size and higher spin rate of any fragments would lower the probability of an interception.

THE LAST DITCH

Nine hours and 36 minutes before SI-4 reached Icarus, the SI-5 interceptor would be launched. Then, as SI-4's warhead exploded at Icarus, SI-6 would be launched. These two interceptors would have a very different mission profile from the earlier launches. By the time they would reach Icarus, the Delta V requirement to deflect Icarus would be too great, even for a 100-megaton warhead. SI-5 and 6 were intended as a last-ditch attempt to fragment Icarus. Fragmenting Icarus could, at best, limit the damage done, as some of the fragments would burn up on entry. At worst, the several large fragments could cause more destruction than an intact Icarus impact. Large fragments could survive entry and, instead of concentrating the damage in one area, could strike all over the world.

The SI-5 interceptor would reach Icarus at about 2:50 p.m. on June 18, 1968. The range at this point was down to 1.41 million miles. The ground radar used to guide the earlier interceptions would now be able to receive the echo from signals reflected from the asteroid. The range data would no longer have to be relayed from the interceptor. If SI-5 was successful in breaking up Icarus, SI-6 would then be targeted on the largest fragment. Otherwise, it would continue on to Icarus. The intercept would be made at about 5:14 p.m. The distance between Earth and Icarus would be only 1.25 million miles. Impact would be 19 hours and 12 minutes away.

As with the earlier interceptors, there were alternate missions for SI-5 and 6. Should SI-1 through 3 be successful in deflecting Icarus, then the launches would be canceled. If SI-4 was successful in completing the deflection, then SI-5 and 6 would be destroyed in space. Should the earlier interceptions break up the asteroid, then SI-5 and 6 would be targeted on the largest fragments.

PROBABILITIES

Of course, Icarus did not strike Earth at 12:26 p.m. on June 19, 1968, and Course 16.74 was only a classroom exercise in "what if." The question remains whether Project Icarus would have been successful in saving Earth, had the effort been real. As with any space mission, the Project Icarus interceptions were an interlinked series of events. The study gave the following estimates of success:

Boosters, 83 percent for all launches
Spacecraft systems, 85 percent for SI-1 through 4, 98 percent for SI-5 and 6
Guidance system, Icarus intact 85 percent for SI-1 through 4, Icarus fragmented 70 percent for SI-1 through 4, Icarus intact 95 percent for SI-5 and 6, Icarus fragmented 90 percent for SI-5 and 6.

For any one interceptor, the odds were 60 percent for a high-altitude shot at Icarus. For a high-altitude shot at a fragment, the odds were 49 percent. For a low-altitude intercept of Icarus or a fragment, the odds were 77 percent and 70 percent. Once past the launch, the odds of success go up. The last two interceptions were the most reliable of all, due to the short ranges and flight times. For both high- and low-altitude interceptions, it is better to go after an intact asteroid than fragments, because of the larger target and slower rotation rate.

As for the probability of a successful deflection (assuming they reached

Icarus and the intercept was successful), the odds for the first four interceptors were

SI-1, 80 percent deflection, 20 percent fragmentation
SI-2, 60 percent deflection, 40 percent fragmentation
SI-3, 43 percent deflection, 57 percent fragmentation
SI-4, 20 percent deflection, 80 percent fragmentation

Clearly, as the range decreases, the odds of deflecting Icarus (and by extension, any asteroid or comet on a collision course with Earth) drop. Closer in, the only possibility is to fragment the target, which can only lower the amount of damage done to Earth, rather than remove the threat entirely. Thus, any such threat must be dealt with at the earliest possible moment, when the Delta-V requirements are smallest. In the final analysis, the odds of a successful deflection were put at 71.4 percent. The probability of fragmenting Icarus and destroying the resulting debris was 19.48 percent. (The specific odds depended on the number of fragments.)[4]

These calculations of probabilities do not take into account the actual events between early 1967 and the spring of 1968. When Course 16.74 began, it seemed Project Apollo was making good progress. In reality, there were deep problems that would take another two years to overcome. The most serious and intractable problem was with the S-II second stage of the Saturn V booster. It was designed to be the most efficient stage ever built. Problems with insulation, welding, cracking of the metal, and North American Aviation's management became apparent soon after the program began. During 1965 and 1966 two S-II stages failed during ground tests. It was not until after the Apollo 204 pad fire, which killed the crew of Virgil I. Grissom, Edward H. White, and Roger B. Chaffee, that the needed management changes were made at North American. Despite being shipped to Cape Kennedy in January 1967, the S-II stage continued to delay the first Saturn V launch.

When Course 16.74 was under way, it was expected that the first Saturn V test flight would be made in August 1967. The need to modify systems and inspect the S-II welds delayed the launch to November 9, 1967. The Apollo 4 mission proved spectacularly successful; the engine flames, billowing exhaust clouds, and Earth-shaking thunder of Saturn V's launch left viewers stunned. The roar drowned out the commentary of CBS correspondent Walter Cronkite and shook loose pieces of the roof and walls of the studio trailer.

The second Saturn V launch, Apollo 6 on April 4, 1968, showed the giant booster still had problems. For the first two minutes of the flight it seemed as flawless as the Apollo 4 launch. Then the booster began to oscillate vertically (called pogo, because it resembled a child bouncing on a pogo stick). Despite this, first-stage separation and S-II ignition were normal. Then, 4.5 minutes into the burn, two of the S-II's engines shut down. The guidance system recomputed the trajectory, and the remaining three engines continued to burn longer than the programmed time.[5] The S-IVB third stage separated and its single engine ignited, but it was both too high and traveling too slowly. To compensate, the S-IVB pitched over toward Earth, then pitched up and in fact was thrusting backward went it reached orbit. It was in an elliptical orbit rather than the planned circular orbit. After two orbits, the command was sent to restart the S-IVB's engine to simulate a high-speed lunar reentry. It failed, and the service propulsion system engine was used to deorbit Apollo 6.[6]

The Apollo 6 launch was the closest any Saturn V mission came to the loss of a booster. The failures were solved with minor modifications, but it was not until August 1968 that the fixes were successfully demonstrated.[7] Apollo 6 was launched only three days before the date of the SI-1 launch. Clearly, it was not enough simply to speed up deliveries of the Saturn Vs. The problems that delayed Apollo 4 and nearly caused the loss of Apollo 6 would have had to be avoided.

THE BOMB

In retrospect, another technical problem that also could have posed difficulties for Project Icarus was the high-yield weapon. The students had assumed that a 100-megaton weapon could be built with no major problems. In fact, the United States would need to conduct atmospheric nuclear tests of a prototype before a production 100-megaton weapon could have been built. In 1963 defense secretary Robert S. McNamara told Congress that a 50-to-60-megaton weapon could be developed for use by B-52 bombers without any testing, while a 35-megaton weapon for the Titan II ICBM required underground testing.[8]

Presumably they represented the limits of existing U.S. high-yield weapons technology in 1967. As with the boosters and interceptors, the development of the weapons was limited by the 70-week deadline, and the weapons would also have to use existing technology. This could have meant that the Icarus interceptors would be restricted to carrying weapons with only a fraction of the planned yield. There was, however, another possibility.

In the autumn of 1954, a little over a year after the Soviets' first H-bomb test, the USSR began a program to develop an air-dropped weapon with a yield of 100 megatons. The weapon was given the code name Ivan and was finally ready for testing in the fall of 1961. During the design process, both the weight and diameter of the weapon had grown. The casing was an elongated teardrop shape, with a cylindrical parachute compartment and three stabilizing fins at the rear. The design changes had made the Ivan weapon too large to fit in the specially modified Tu-95V's bomb bay, and it had to be carried externally. On October 30, 1961, the Tu-95V took off and flew to the Soviet nuclear test site at Novaya Zemlya Island. Over the target area, Ivan was dropped from the bomber and exploded with a yield of 56 megatons, the largest manmade explosion ever. A production version of Ivan could be weaponized as a 100-megaton bomb.[9]

Subsonic bombers like the Tu-95V were vulnerable to air defenses, however. On April 29, 1962, Soviet premier Nikita Khrushchev approved development of the UR-500 missile. It was to serve four different roles: as a very large ICBM to deliver the 100-megaton weapon, as the GR-2 global rocket to place large nuclear warheads in orbit, as a launch vehicle for large military payloads, and as the booster for a Soviet manned lunar flyby mission.[10]

The ouster of Khrushchev in October 1964 ended the UR-500's role as a ballistic missile. But the superbomb program meant that by the mid-1960s the Soviets would have had the design data needed to produce a small number of 100-megaton nuclear weapons of the type envisioned by Project Icarus. Just as the United States lacked the weapons needed, the Soviets lacked the booster. The Soviet N 1 Moon rocket did not make its first launch attempt until February 21, 1969. It failed after 70 seconds. Three more N 1 launches were made, on July 3, 1969, June 27, 1971, and November 23, 1972. Each time, the boosters failed before first-stage separation.[11]

With neither side having a complete defense and the whole world at risk, the United States and the Soviet Union would have been forced to put aside their differences. The problems of such cooperation should not be underestimated. In 1967 the Soviet Union was a closed society, and the existence of a Soviet Moon landing program was a state secret. Yet for Project Icarus to work, the Soviets would have to turn over to the United States six 100-megaton nuclear weapons, the biggest state secret of all.

AFTERMATH

Three decades after Course 16.74 first met, it is clear that Project Icarus was a landmark study. It described the basics of a planetary defense system and pointed out the requirements it would have to meet. John Rather echoed many of the same issues in his March 24, 1993, statement to Congress: "Early detection obviously gives a much longer reaction time. More importantly, interceptions far from Earth—made feasible by early warning—are much more desirable and easier than interceptions near Earth."

Early detection meant that even a relatively small Delta-V change far from Earth would result in a large miss distance. Another advantage was that if the asteroid or comet fragmented in the interception, then all the fragments would miss Earth. This was essential, Rather noted, "because we don't want to convert a cannon shell into a cluster bomb." Finally, early detection allows a shoot-look-shoot strategy, which would maximize the chances of a successful deflection. This involves firing an interceptor at the target, measuring the deflection, and then, if needed, launching another interceptor.

Rather also noted that early warning made mission planning much easier. If an impact was predicted several decades or more in advance, then precursor missions could be launched to sample the asteroid and provide detailed information needed for the deflection. The MIT students had to guess the size and composition of Icarus, which affected the mass estimates and the resulting Delta V of the deflection attempt. The interceptors could also be launched at a time that would minimize the energy required and maximize the effectiveness of the Delta V. If the warning time was on the scale of only a few years, however, the problem became more difficult. Precursor missions would be harder to accomplish, meaning the deflection might have to be attempted with major unknowns. The interceptor launch energy requirements could also be higher.

If the warning time was less than a year, as it might be for a large long-period comet, the problems would be far greater. There would be no chance for precursor missions, and the interceptors would have to be launched in short order. The launch energy requirements would be much higher, possibly as much as 100 times that of the first case. This situation is similar to that facing Project Icarus. The short development time and the 60-day limit on the interceptor flight time meant that Icarus had to be deflected relatively close to Earth, which would require a 100-megaton warhead, which in turn required a Saturn V to carry the weapon. The final case is an impact with only a few days of warning. Rather noted, "There is at present no response that has a high probability of success." With an improved detection system, however, that case need never arise.[12]

Since the Project Icarus study, there have been many changes in both space technology and American society. At century's end, the three largest boosters available were the U.S. Titan 4B, the Russian Proton (developed from the UR-500 ICBM), and the French Ariane 5. Although they are all smaller than the Saturn V, other advances may compensate. Space technology is now much more reliable than in the late 1960s. Voyager spacecraft have operated for more than two decades, and the multiplanet gravitational assist profile has been used on a number of space probes. It allows a booster to launch a payload much too heavy for a direct flight to the target asteroid by using the gravitational attraction of other planets for added velocity. The price, however, is a much longer flight time. This option would not be available if warning time is short.

Although nuclear weapons remain the only feasible means to deflect an asteroid, the attack profile has been greatly modified over that used in Project Icarus. Rather than a single very large weapon, smaller devices are now seen as both practical and necessary to prevent fragmentation. Johndale Solem, a mathematical physicist at Los Alamos, has proposed that a neutron bomb be used. It would release most of its energy in the form of neutrons, high-energy particles that would penetrate into the asteroid, then blast away the surface layer.

Solem determined that the optimum burst height above the surface would be about one-half the NEA's radius, providing the optimum Delta V with the least chance of fragmenting the object. For a 1-kilometer asteroid, a weapon with a yield of 4 megatons would be required to produce a Delta V of 0.5 meter per second. With that relatively low Delta V, the asteroid would have to be intercepted at least five months before impact to miss Earth. With more warning time, much smaller weapons could be used. Thomas Ahrens of Caltech and Alan Harris of JPL calculated that a 100-kiloton explosion could produce a Delta V of 1 centimeter per second in a 1-kilometer asteroid. If done years before the impact was to occur, even this tiny velocity change would be sufficient.

Another approach was proposed by Anthony Zuppero of the Idaho National Engineering Laboratory, based on a series of one-two punches, rather than a single bomb. The first bomb would explode, blistering the surface and raising a cloud of dust. A second bomb would follow close behind and then explode, vaporizing the dust cloud and giving an added push. It had the advantage of further reducing the chance of fragmentation. Again, early warning is critical. As the asteroid or comet nears Earth, the Delta-V requirement increases at an accelerating rate.[13]

There has also been a reassessment of the threat posed by smaller objects. They had been dismissed in the Spaceguard study as unimportant, as they would cause only local or regional damage. Some astronomers now believe that the impact of objects as small as 0.2 kilometer would have a devastating social effect. Alain Maury, a French asteroid hunter, observed, "Even without an impact winter, after the initial fatalities, our banking system goes berserk, all multinational companies—many of them dealing with food distribution—do the same, and we do not find our cereal in our supermarkets anymore."

Michael Baillie of Queen's University in Belfast estimated a 0.5-kilometer asteroid would be sufficient to end modern civilization. "The trouble is," he noted, "that a significantly smaller impact could still do the trick, especially if over an ocean. . . . Civilization is a thin veneer. Take away all air travel, restrict global food supplies, demonstrate that the military and governments are ineffective, demonstrate that coastal zones should be avoided, and where would we be?"[14]

As Baillie noted, this new attitude is based on a realization of the effects of an ocean impact. Computer models by Jack G. Hills and Charles Mader of Los Alamos National Laboratory looked at the tsunamis created by asteroids of various sizes. The worst case was the impact of a 5-kilometer asteroid in the mid-Atlantic. Within three hours, tsunamis would hit the Eastern Seaboard of the United States from New England to the Carolinas. In the northeast, the tsunamis would sweep inland to the foothills of the Appalachian Mountains. Not only would New York

City be destroyed but also Washington, D.C., Philadelphia, Baltimore, and Richmond. Coastlines in Europe, such as Ireland, France, Portugal, and Spain, would also be wiped out. A similar impact in the Pacific Ocean would flood the entire Los Angeles basin.

An impact of this scale would occur at intervals of roughly 10 million years, but Hills noted, "Any asteroid over 600 feet in diameter, we have a real problem." Such smaller impacts are 2,000 to 3,000 times more likely.[15] Even a 5-megaton impact would generate tsunamis comparable in size to those caused by the largest earthquakes. Impacts on this scale occur at intervals of centuries, rather than millions of years.

Yet these projections and computer models remained only theoretical possibilities. Despite S-L 9, the human belief that it can't happen here was intact. And then, for one day, it stopped being a theoretical possibility.

1997 XF11

On December 6, 1997, Jim Scotti was searching for two returning comets with the Spacewatch telescope. The night was poor, with off-and-on clouds and bad seeing. The weather forecast also predicted rain for the final two nights of the run. The first nine objects spotted by the automatic search program were normal main-belt asteroids. The tenth was an NEA, which was later designated 1997 XF11. During the next two weeks, observations from two Japanese amateur astronomers allowed an orbit to be calculated. The object was in a low-inclination orbit with a period of 1.73 years. Given its distance and brightness, the asteroid was estimated to be between 1.2 and 2.4 kilometers in diameter. It was the one-hundred-sixtieth NEA discovered at Spacewatch and the fifty-eighth by Scotti.[16]

This initial orbit also indicated that the minimum distance between the orbits of Earth and 1997 XF11 was very small, and it became the one-hundred-eighth PHA. As more positional data came in, it became apparent that 1997 XF11 would pass very close to Earth in October 2028. An orbit calculated using 60 days of observations indicated that the miss distance would be 500,000 miles, twice the distance between Earth and the Moon. Additional observations were made on March 3 and 4, 1998, by Peter Shelus with a 30-inch telescope at McDonald Observatory in western Texas. The observed arc of 1997 XF11 was now extended to 88 days.

The new orbital computations indicated a miss distance of only 30,000 miles from the center of Earth (which is about 4,000 miles in radius). The calculated time of 1997 XF11's closest approach was around 1:30 p.m. EDT on Thursday, October 26, 2028.[17]

Because this 88-day arc represented only a small part of 1997 XF11's orbit and the calculated orbit was then being projected a full 30 years into the future, there was a sizable margin of error, amounting to roughly 112,000 miles. Given the area of this error, compared to the surface area of Earth, there initially appeared to be roughly a 1 in 900 chance of an impact.[18] If it did hit, however, the explosive yield would be on the order of 500,000 megatons, the same as the estimate for Icarus in the 1967 study. A land impact would shroud Earth in debris for three to five years, and an ocean impact would leave coastal cities as rubble-strewn mud flats.[19]

On March 11, 1998, Brian Marsden issued a Minor Planet Electronic Circular on 1997 XF11. The announcement five years before that comet Shoemaker-Levy 9 was going to hit Jupiter was very low-key. The 1997 XF11 circular was more direct. After giving details of the discovery, it said, "This nominal orbit indicates that the object will pass only 0.00031 AU from the Earth on 2028 October 26.73 UT! Error estimates suggest that passage within 0.002 AU is virtually certain, this figure being decidedly smaller than has been reliably predicted for generally fainter PHAs in the foreseeable future." This was the first time ever that an exclamation point had appeared in a Minor Planet circular. After noting that further observations were required to refine the orbit, Marsden asked that a search be made for prediscovery photos of the asteroid. Earlier close approaches had taken place in 1990, 1983, 1976, 1971, and 1957.

Eleanor Helin was among those who received the e-mail circular. She was sure that photos from her Planet-Crossing Asteroid Survey would show 1997 XF11. The next morning, archivist Kenneth Lawrence began searching photos taken on March 22 and 23, 1990. After several hours, he located 1997 XF11 as a very faint dot. Its position was measured and sent to Marsden. With data covering eight years, it was possible to erase any uncertainty. That evening Marsden, Donald Yeomans, and Paul Chodas of JPL announced nearly identical predictions for 1997 XF11's future orbit. The October 2028 miss distance would be no less than 600,000 miles, more than twice the distance to the Moon.[20]

The most significant outcome of the 1997 XF11 episode was on an emotional level. For a brief time, people contemplated not merely their personal mortality but that of humanity itself. Although personal mortality is inevitable, one hopes that humanity will continue. The 1997 XF11 episode showed how fragile that belief is. Everything humanity had accomplished in art and science could, in a single flash of light, vanish. All that humanity had worked and fought for, every noble deed, every sacrifice, every hope for a better tomorrow, would have been in vain.

THE TORINO SCALE

In the wake of 1997 XF11, it was clear that procedures had to be established regarding threat predictions. The fear of an impact occurred because it had been assumed that the error ellipse was spherical. In fact, it was shaped more like a straight line, one that never touched Earth. Looking to the future, with several asteroid search programs under way, it was inevitable that there would be more predictions of close approaches. It was judged important by both NASA and astronomers to avoid a series of false alarms.

Unfortunately, such a false alarm followed 1999 AN10's discovery by the LINEAR telescope on January 13, 1999. A group of Italian astronomers calculated that the asteroid would make a close pass in 2027 and that there was a one-in-a-billion chance of an impact in 2039. When their results (which, like 1997 XF11, were based on limited data projected four decades into the future) were prematurely released by outside individuals, there was a flurry of press stories. Additional observations allowed a refined orbit, which indicated that 1999 AN10 may come as close as 30,000 kilometers in August 2027. The margin of error was still high, and it was possible that the effects of Earth's gravity during the pass could put the asteroid on a trajectory resulting in a collision in 2044 or 2046.

Then two German amateur astronomers, Arno Gnadig and Andreas Doppler, found an image of the asteroid on a 1955 photo taken by the 48-inch Schmidt. Another 48-inch Schmidt image showing 1999 AN10 was found by Gareth V. Williams of the Minor Planet Center. Williams and Brian Marsden measured its position on the photos and calculated that it would come no closer to Earth than 386,000 kilometers in 2027 and that an impact was impossible in 2044, 2046, or for several decades after.[21]

Richard P. Binzel was among those disturbed by the sensationalism surrounding 1997 XF11 and 1999 AN10. He said later, "I was incredibly frustrated that we didn't have a hazard index at our disposal during the XF11 affair." He had suggested a 0-to-5 scale in 1995, but it attracted little interest. In early 1999 Binzel realized that combining an expanded 0-to-10 scale with color would be the solution. The similarity with the Richter scale for earthquake measurement was also an advantage, as it would make it acceptable to both press and public. Binzel's scale (shown in table 2) was adopted by the IAU in July 1999 and named for the Italian town of Torino, where a workshop on impacts was held. Under the Torino scale, both 1997 XF11 and 1999 AN10 would have been classified as a 1 following their discovery and then would have been quickly downgraded to a 0. "Imagine," Binzel said, "if this scale had been in place two years ago."[22] The impacts that caused Meteor Crater and Tunguska would have been an 8 on the Torino scale.

PLANETARY DEFENSE AND THREAT ASSESSMENTS

The events surrounding the 1997 XF11 prediction also brought renewed attention to planetary defense. In his congressional testimony, Rather noted the uncertainties affecting any decision to develop such a planetary defense. "Whether it is safe," he said, "to defer action for fifty years while we await a more robust, higher technology space program is anyone's guess." There was a belief by some of the workshop members that any asteroid on a collision course with Earth would be detected several decades in advance. Any development and testing of a planetary defense, they believed, should be deferred until an impact was actually threatening. Rather concluded, "There is a clear need for continuing national and international scientific investigation and political leadership to establish a successful and broadly acceptable policy."[23]

Although the technological problems of building a planetary defense are understood, how a "broadly acceptable policy" could be developed is unclear. The social climate today is very different from that of three decades ago. When the Project Icarus study was completed in 1967, the press response was positive. The 1992 Interception Workshop met with the opposite reception. The news media universally lambasted it, depict-

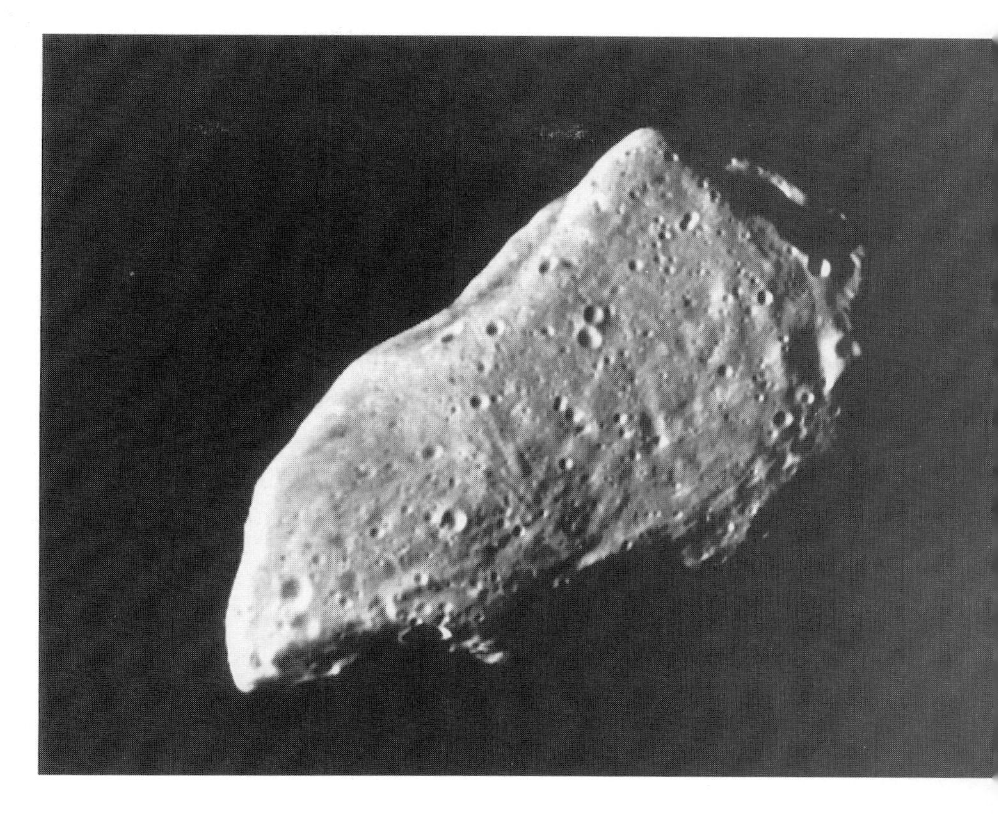

Asteroid 951 Gaspra, photographed by the Galileo spacecraft on October 29, 1991. This was the first close-up look at an asteroid. (NASA photo)

A B O V E : Asteroid 243 Ida and its moon Dactyl from the Galileo spacecraft, August 28, 1993. Despite numerous attempts by ground-based observers, this was the first proof that asteroids did have moons. (NASA photo)
T O P O P P O S I T E : Asteroid 4 Vesta as viewed by the Hubble space telescope in May 1996. The two top images are the actual Hubble image (left) and a 3-D computer model (right) constructed from the Hubble data. Both show Vesta's 13-kilometer-high peak at the bottom. The lower center image is an elevation map showing the 640-kilometer impact basin and the peak as a bull's eye on the lower half of the asteroid. (NASA photo; computer model image by Ben Zellner and Peter Thomas) B O T T O M O P P O S I T E : Ground-based infrared photographs of the S-L 9 fragment R impact on Jupiter. The image on the right was taken 2 minutes after that on the left. They show the brightening as the fireball, on the left edge of Jupiter, grew in size. (NASA photo by Yongha Kim, Beth Clark, and William Cochran)

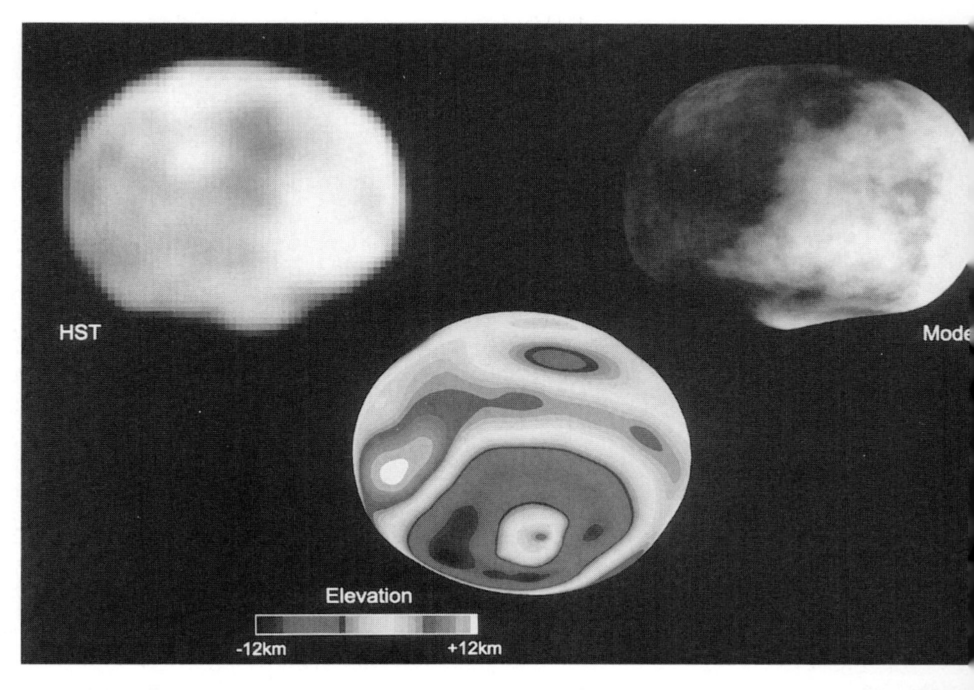
HST
Mode
Elevation
-12km
+12km

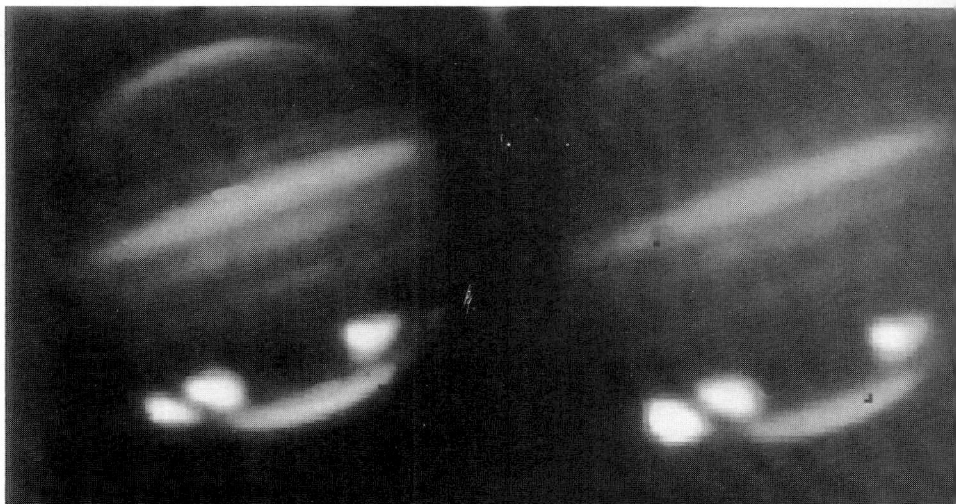

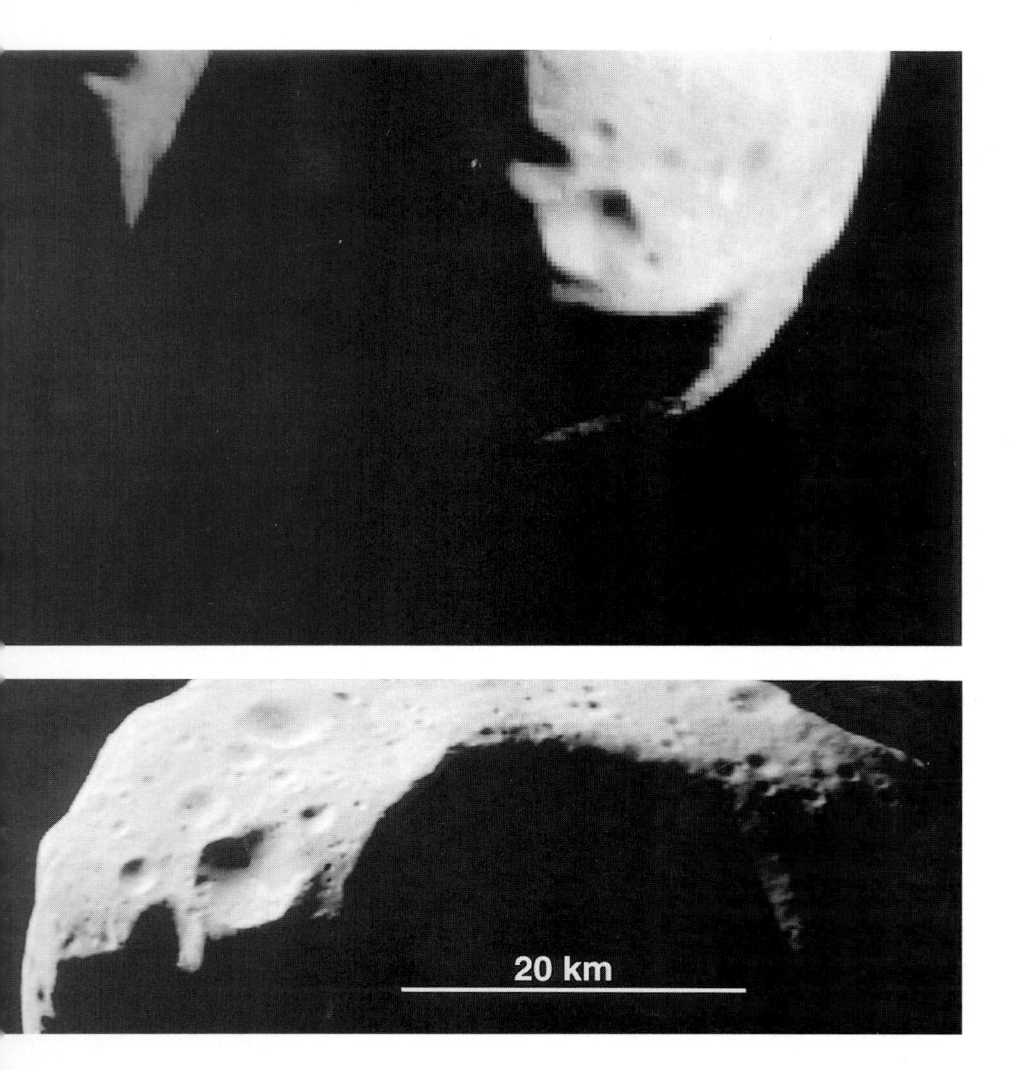

TOP: The first NEAR image of asteroid 253 Mathilde, taken on June 27, 1997. The large dark area on the left is believed to be an impact crater more than 19 kilometers deep. BOTTOM: A portion of the battered surface of 253 Mathilde. The asteroid shows numerous large craters from massive impacts; at least five of its craters are larger than 20 kilometers in diameter. (Johns Hopkins University Applied Physics Laboratory photos)

An asteroid family portrait: Mathilde, Gaspra, and Ida as viewed by NEAR and Galileo. All the images are to the same scale and show the asteroids' relative size. (Johns Hopkins University Applied Physics Laboratory photo)

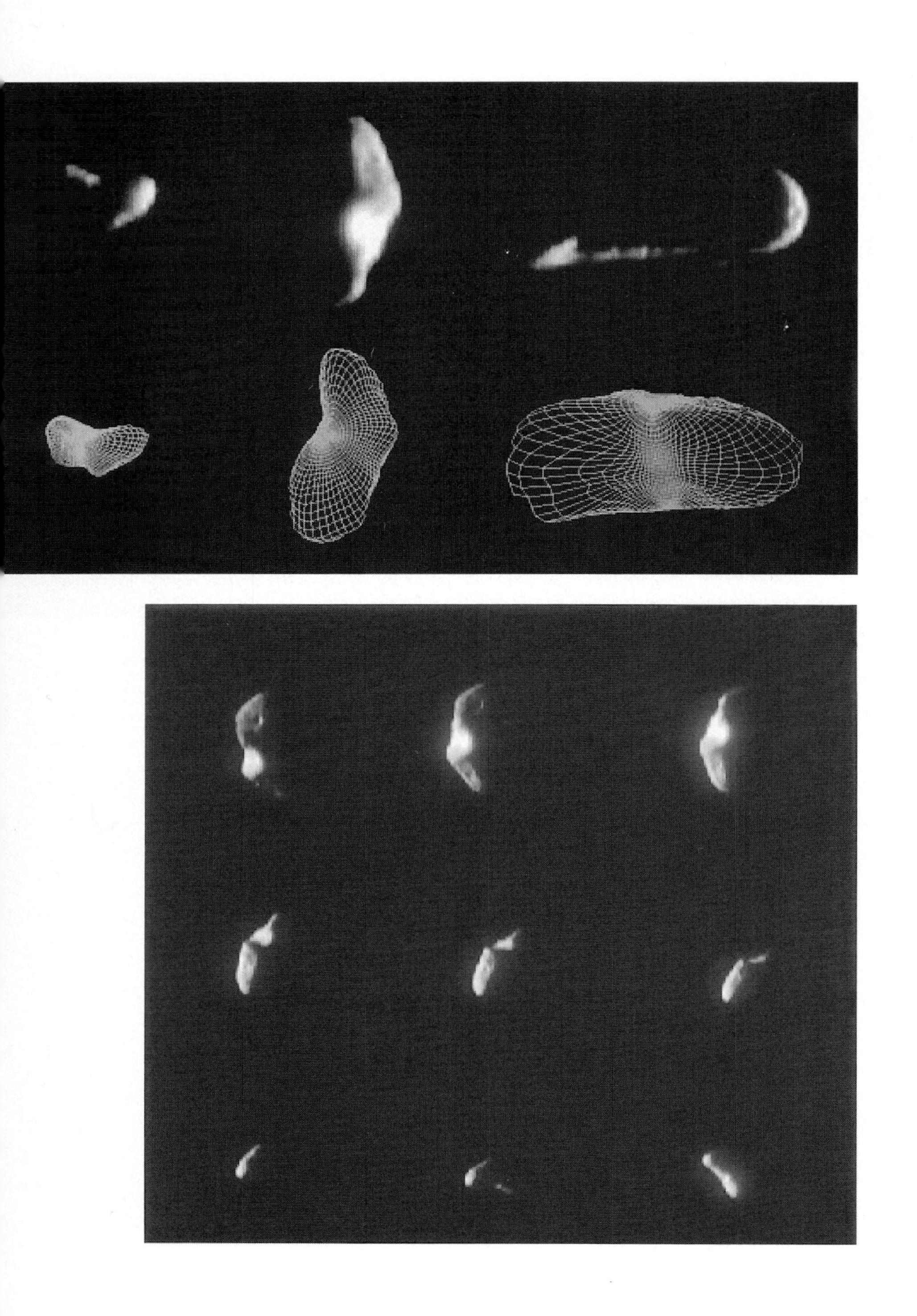

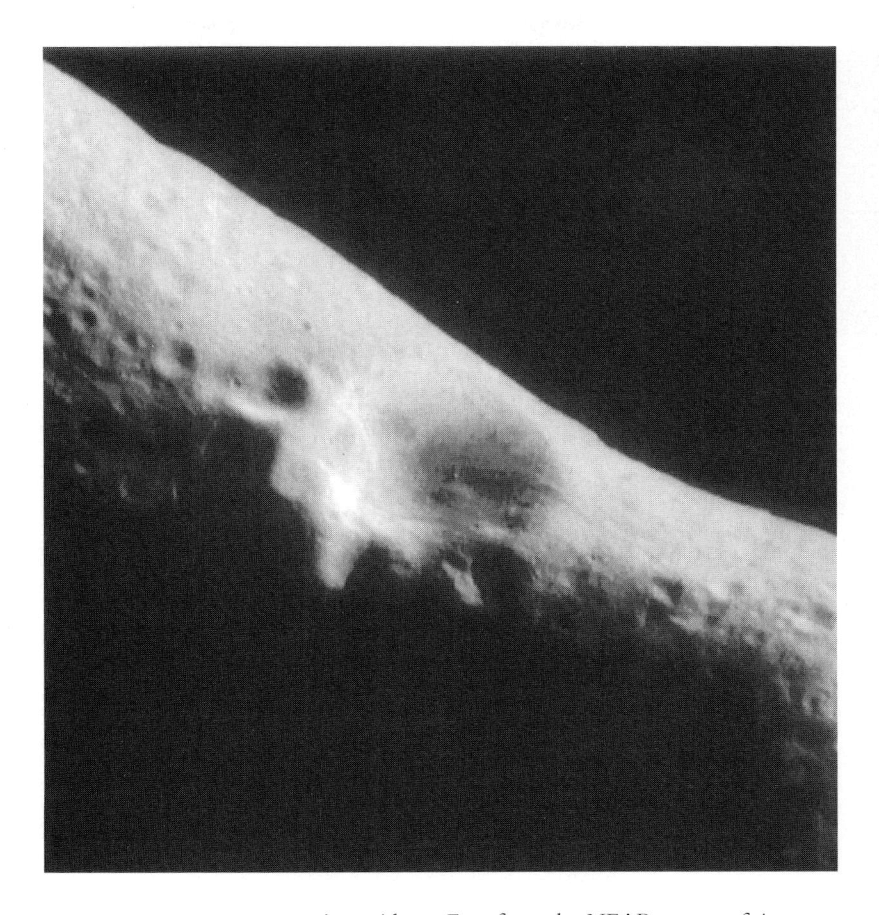

TOP OPPOSITE: Asteroid 433 Eros from the NEAR spacecraft in December 1998. An engine problem prevented the spacecraft from orbiting Eros, but as it flew by, it was able to get long-range images (top row). Below them are computer-generated models of the asteroid's shape. BOTTOM OPPOSITE: A series of images of 433 Eros as the NEAR spacecraft flew by on December 23, 1998. The images show the asteroid's rotation as well as its changing position relative to the NEAR spacecraft. ABOVE: The first photo of 433 Eros taken by NEAR following its orbital insertion. The initial analysis of the images showed a varied surface that has been heavily cratered. (Johns Hopkins University Applied Physics Laboratory photos)

TOP: Comet Shoemaker-Levy 9, as seen by the Hubble space telescope in May 1994. The comet's 21 fragments stretched across a distance equivalent to three times the distance from Earth to the Moon. (NASA photo)
BOTTOM: Dr. Eugene Shoemaker. (California Institute of Technology photo)

Table 2 THE TORINO SCALE

Color	Rank	Threat Level	Consequences
White	0	No likely consequences	Likelihood of a collision is zero, or well below the chance that a random object of the same size will hit Earth within the next few decades.
Green	1	Merits careful monitoring	Collision extremely unlikely.
Yellow	2	Merits concern	Close encounter—collision very unlikely.
	3		Close encounter with a 1-percent or greater chance of impact capable of causing localized destruction.
	4		Close encounter with a 1-percent or greater chance of impact capable of causing regional devastation.
Orange	5	Threatening events	Close encounter with a significant threat of an impact capable of causing regional devastation.
	6		Close encounter with a significant threat of an impact capable of causing a global catastrophe.
	7		Close encounter with an extremely significant threat of an impact capable of causing a global catastrophe.
Red	8	Certain collisions	A collision capable of causing localized destruction.
	9		A collision capable of causing regional destruction.
	10		A collision capable of causing global climatic catastrophe.

ing the participants as nuclear madmen intent on preserving their Cold War empires. The possibility of asteroid impact was dismissed as a fraud, an invented threat devised to justify continued unnecessary military spending.

A concerted search effort could give two or three decades of warning, but the case of 1997 XF11 and subsequent episodes indicate this may not be as clear-cut as previously thought. There had always been an unspoken assumption that it would be certain that an asteroid was on a collision course with Earth. It is now apparent that astronomers would initially be able to say only that there was a significant chance of impact. A final determination may take years.

Scotti gave one possible scenario in which an NEA is spotted and observed for two months. The initial orbit indicates it will make a close pass in 30 years. The error ellipse measures 2.8 million kilometers long but only a few thousand kilometers wide. This time, Earth does lie within the error ellipse, giving a 1-in-200 chance of an impact. An archive search turns up several photos taken 8 years before, which narrows the error ellipse to 175,000 kilometers long and 1,000 kilometers wide. It still includes Earth, making the odds of a collision 1 in 8. The asteroid is now too far away and dim to make new observations, so astronomers must wait 4 years until it is properly placed to track it with ground-based radar. This will give the final verdict as to whether it will impact 26 years later.

But, as Scotti noted, this is only one possibility. There might be no archive photos, in which case several years would pass before the orbit could be improved. Another was that there would not be a close pass for a decade or more after it was detected. To put the odds in perspective, Scotti noted that a 1-in-1,000 chance of a collision is the same probability that a no-hitter will occur in any specific baseball game.[24]

There is no chance that such a possible future impact could be kept secret for a prolonged period of time. Astronomical observatories are not secure facilities, and any such effort would involve astronomers, graduate students, staff members, and archivists, as well as heavy telephone and e-mail traffic. The news could not be kept hidden, certainly not for four years. How would society react to the possibility of a future impact? Would people panic, as some suggest, or would this otherworld threat unify mankind, as others propose? Or would even the possible end of the world fail to overcome politics-as-usual?

Several months before the S-L 9 impact occurred, John Boudreau made a sarcastic observation in a *Washington Post* article. He said that while a search program could give 20 or 30 years of warning, all this meant was that there was "Time enough for the usual factions to form." He wrote, "There'd be the Universe Firsters, who believe it's wrong to intervene," as well as bumper stickers saying "Save the Asteroids" and "Asteroids Were Here First." He also predicted the impact would be used in the blame game: "Feminists would blame men. Rush Limbaugh would blame Whitewater. Californians might worship the thing."[25]

Sarcasm aside, due to the low probability of an impact, the giggle factor, and the ambiguous nature of the possible warning, any attempt to develop a planetary defense is sure to bring political and academic condemnation, tabloid stories of cover-ups and failures, and lawsuits attempting to stop it. Political scientists can be expected to write learned analyses, calling development of a planetary defense destabilizing and accusing it of having a harmful effect on arms control. There is, however, one other factor.

HOW DECISIONS ARE REALLY MADE

This is not the first time a threat from space has been the subject of political controversy. After the launch of Sputnik, President Dwight Eisenhower and his scientific advisers dismissed public fears that the Soviets would attempt to place nuclear weapons in orbit. They believed that it made no military or technical sense. This opinion continued to guide policy under John F. Kennedy. It was assumed that the Soviets would realize the futility of orbital nuclear weapons and abstain from developing them. As noted earlier, however, there was a Soviet orbital nuclear weapons development program under way in the early 1960s.

Kennedy's advisers also assumed that the Soviets would not put nuclear-armed ballistic missiles into Cuba, as that would, like orbital nuclear weapons, be irrational. Following the Cuban missile crisis, when the Soviets did put missiles into Cuba and brought the world to the brink of nuclear war, the question of orbital nuclear weapons was reconsidered. To meet the possibility, the Air Force proposed Program 437, a Thor ballistic missile armed with a nuclear warhead. The Thor would be launched toward the target satellite, then, at a precalculated point, the warhead would be triggered by a radio command.

In late 1963 Kennedy Administration officials met to discuss the technical feasibility and political sensitivity of Program 437. On hand were the secretary of defense Robert S. McNamara, the director of defense research and engineering Harold Brown, undersecretary of state U. Alexis Johnson, the director of the U.S. Information Agency Edward R. Murrow, and the chief of research and development of the Joint Chiefs of Staff Col. Harry Evans. Despite the increased concern about the possibility of Soviet orbital nuclear weapons, the senior civilian leadership of the Defense and State departments were very nervous about Program 437. They especially did not like the idea of a nuclear explosion in space, and the discussion seemed to be going against what was seen as a major political liability.

Murrow had been quietly smoking a cigarette. He interrupted with a short remark: "If the Soviets place a bomb in orbit and threaten us and if this administration has refused to develop a capability to destroy it in orbit, you will see the first impeachment proceeding of an American President since Andrew Johnson."

About two minutes of total silence followed Murrow's remark. Finally, McNamara said testily, "Well, it doesn't cost much, and the JCS want it, so let's approve 437." Kennedy Administration officials viewed Program 437 as a political menace, but not having it was seen as political suicide, and that was the more important consideration. Program 437 became operational on May 29, 1964, and was not retired until April 1, 1975.[26]

An asteroid or comet impact is the only natural disaster that can wipe out human society and the only natural disaster that human society can prevent. A large impact is an improbable event that is absolutely guaranteed to occur. Over the span of geological time, very large impacts have happened countless times and will occur countless more times in ages to come. Yet in any given year, or in one person's lifetime, the chance of a large impact is vanishingly small. The same, it should be noted, was also true when the dinosaurs ruled Earth. Then, on one ordinary day, probability arrived in the form of a comet, and their world ended. Their only tombstone was a thin layer of iridium.

As can be ours.

12

THE THIRD CENTURY OF ASTEROID STUDIES

listen: there's a hell of a good universe next door; let's go.

E. E. Cummings, "pity this busy monster, manunkind"

THE NEW MILLENNIUM ALSO marks the beginning of the third century of asteroid studies. As we prepare to embark on that unknown voyage, it is a fitting time to look back, to find some guideposts for the journey. The history of the past two centuries offers insights into the process of discovery, the changes in both asteroid studies and science over this period, the politics of science, and science versus politics.

THE DISCOVERERS

One aspect that stands out is the unusual background of many of those involved in asteroid studies. Giuseppe Piazzi, the discoverer of Ceres, had a history of unconventional thinking, which added to the uncertainties of his academic life. More important was his outlook. When Piazzi noticed that a dim star had moved, he was able to recognize the possibilities it represented. It is clear that he was thinking the object was the missing planet very soon after his discovery.[1]

The same was also true for Baron Franz von Zach. Today, academic journals are the norm in the physical and social sciences. They provide the basic means of communication between specialists in a particular field. It was Zach who first came up with the idea. It was also Zach who, it can be argued, originated the idea of big science. The Celestial Police was a cooperative effort of astronomers working toward a common goal. Each was assigned a particular part of the sky to search for the missing planet. Zach realized that the search program he envisioned was too large to be undertaken by a single astronomer; it required a team effort.[2]

We can see this in our own time as well, with the stories of Gene Shoemaker and Walter Alvarez. In 1948 Shoemaker realized what few at the time could, that a manned flight to the Moon was a real near-term possibility. During his later work in the U.S. nuclear weapons program, he witnessed the power involved in an impact and could compare the effects the blasts produced to those in Meteor Crater. Although Shoemaker was trained as a geologist, he became an astronomer to undertake the asteroid search program. Shoemaker was a success in both fields and brought a crossover of ideas and insights to what had been separate specialties.[3]

With Walter Alvarez, one can see the influence of his father, Luis, whose activities spanned radar, nuclear physics, UFOs, Chephren's Pyramid, and the 1968 Nobel Prize. Given the background of his father, it is not surprising that Walter Alvarez was willing to consider sources of the iridium layer that would not have occurred to a geologist with a more conventional background.[4]

That is the underlying importance of this pattern. Discoveries are made by those who are mentally prepared when the opportunity presents itself. They are more willing to think out of the box and willing to propose ideas that do not follow existing patterns. The reason is that individuals with wider varieties of backgrounds and experiences are more likely to see connections that would be missed by individuals who have not had such experiences. It also implies a willingness to take risks, rather than stay on the standard career track. Of course, that is no guarantee one is right. Most bright ideas prove to be dead ends rather than breakthroughs, and working outside one's specialty carries risks. At the same time, the only people who never run the risk of failure also never try.

CHANGES IN ASTEROID STUDIES, 1801–2001

The past three decades have brought a fundamental change in asteroid studies. For the first time we have been able to learn the basic physical properties of asteroids, such as rotation rate, size, shape, composition, and origin. This stands in contrast to the first 170 years of asteroid studies. For most of that time, all astronomers could do was estimate their brightness and calculate their orbits. It was not until 1894 that the first size estimates were made, 1901 that the first light variations caused by rotation were confirmed, and 1918 that the existence of asteroid families was put on a solid basis.

It was the twin technological revolutions of electronics and computers that made this change possible. Previously, improvements had been incremental—the photographic techniques used by Max Wolf in 1891 were similar to those used by the Shoemakers in the early 1990s. In contrast, the automated computer search of LINEAR, NEAT, and Spacewatch represents a technological breakthrough in terms of speed and precision. In the same way, development of first photomultiplier tubes and then CCDs provided the means to determine asteroid composition. With radar astronomy, advanced technology and computer processing have allowed us to image asteroids directly.

As a result, the process of asteroid research has gone in two different directions. The changes in technology have put a premium on teamwork. The research cuts across different scientific specialities and requires not only astronomers but also geologists, chemists, radar and optical engineers, and computer software designers. This also puts a premium on project management, of the different specialities and between different organizations, such as NASA, JPL, the Air Force Space Command, and the National Reconnaissance Office.

Yet these advances in technology also mean that the amateur once again has a role in asteroid studies. CCD cameras are advertised in astronomy magazines, and personal computers can process images. It is now within the means of an amateur to undertake an asteroid search program. Although a backyard telescope will not have the light-gathering power of LINEAR, NEAT, or Spacewatch, the success of Roy Tucker in the first discovery of an Aten asteroid by an amateur illustrates the possibilities.[5] More important, the likelihood of finding an NEA is dependent on the number of telescopes searching.

As significant as the achievements of the past three decades have been, the launch of the first asteroid space missions heralds an even greater understanding. Any Earth-based observations are, by necessity, at a distance. A space mission can fly by an asteroid at close range or go into orbit for prolonged study. A landing permits direct examination of the asteroid's surface material and structure, as well as the possibility of returning samples for analysis on Earth. Such observations can then also be used as a calibration of Earth-based observations.

THE POLITICS OF SCIENCE

The past two centuries of asteroid studies also provide examples of the politics of science, that is, how scientific ideas are formulated, how they come to be accepted or rejected, and the interplay between individual scientists and specialties.

In the question of the extinction of the dinosaurs, an unusual iridium layer was discovered that seemed, upon examination, to be from an impact. The geological layers above and below it provided evidence that this had caused mass extinction, while similar evidence at other sites indicated it was not a local event but rather a worldwide occurrence. Walter and Luis Alvarez concluded that the dust blasted into the atmosphere by the impact blotted out the Sun and caused the extinction. They published their evidence and the conclusions based on it.

In response, other scientists then offered counterarguments or supporting evidence. In the process, the idea was refined; rather than a simple cloud of dust, evidence was uncovered for a much more complex mechanism involving darkness, acid rain, and periods of cold and heat. As the debate continued, the weight of evidence favoring the impact theory built up, and the arguments of those opposing the theory were increasingly unable to counter it. Finally, with the discovery of the tsunami deposits and the crater itself, scientific opinion shifted to acceptance of the impact extinction.[6]

In the case of Vulcan, the reverse was true. The theory of an inner asteroid belt was proposed by Urbain J. J. Leverrier to explain irregularities in the orbit of Mercury that were outside classical gravitational theory. Vulcan was, in effect, an effort to preserve the status quo of Newtonian physics. Sightings of objects were interpreted as bodies crossing the Sun.

Based on these sightings, predictions were made of future transits of the Sun by Vulcan. But here the problems appeared: Vulcan was not seen at the predicted times. No patterns appeared among the sightings, and other observers looking at the Sun at the same time saw nothing.

Critics of Vulcan, among them C. H. F. Peters, argued that the sightings were errors, such as mistaking two small sunspots for a single moving object. The repeated failure to find evidence of Vulcan led to the eventual rejection of the theory. In this case, it was the failure to find evidence, when that evidence should be found, and the failure of the predicted transits to occur that led to the rejection of the theory. It was not until Einstein's general relativity theory of gravity that the orbital changes of Mercury were explained. The explanation of these irregularities did much to aid in the acceptance of Einstein's theory.[7]

Beyond the antiseptic picture of the making of theories and the collection of data, there is also a human dimension. It includes personal differences, emotional commitment, conflicting scientific outlooks, turf battles, and political influences. Leverrier was a mathematical astronomer; for him astronomy began and ended with calculations on paper. Peters was an observational astronomer who spent his days observing the Sun and his nights searching for asteroids. Leverrier was a French national hero for the calculations that led to the discovery of Neptune, while Peters had endured a lifetime of hardship. Perhaps these differences were responsible for Peters's contemptuous attitude toward Vulcan.

In the case of the extinction of the dinosaurs, a similar set of conflicts operated. The paleontologists who were critics of the theory seemed to resent the interlopers who had proposed it. The paleontologists saw the question solely through the prism of the fossil record and the doctrine of uniformitarianism. Impact studies, the Apollo flights, and planetary missions were outside their field and therefore bad science. Besides turf battles, there was also a difference in outlook. Geologists and paleontologists looked to the rocks at their feet for scientific answers; the earth was the center of their scientific universe. For astronomers, scientific answers are found by searching the solar system and the universe beyond.

On an individual level, an influencing factor is emotional commitment to a scientific position. Because of this, no matter how much evidence is assembled to support or refute a theory, there will always be some scientists who will never accept it. Leverrier's belief in Vulcan never wavered, nor did that of his American supporters James C. Watson and Lewis

Swift. A few geologists, even after Shoemaker's research and the Apollo landings, still believed that craters were caused by volcanoes.[8] One of the author's geology professors in 1979 did not accept plate tectonics. He repeatedly dismissed it by saying the evidence was not sufficient. He added, however, that he would be retiring at the end of the semester and the students would not have to put up with his unpopular opinion for much longer.

One can see how emotional commitment enters into scientific theories in regard to asteroid impacts and impact extinctions. Many of America's most eminent geologists believed that Meteor Crater was a volcano and that the craters on the Moon had a similar cause. The notion of impact extinctions challenged the basic idea of geology, that all geological processes were slow and gradual. Any theory that proposes that the underpinnings of one's scientific research are not valid will meet resistance.

A significant change in recent decades is the intrusion of politics into the scientific process. The question of Vulcan's existence had no bearing on politics in the United States after the Civil War. In contrast, issues such as the nuclear freeze, nuclear winter, global warming, and SDI became linked to the impact extinction theory. In the second half of the nineteenth century, the advice of scientists was not sought on public policy issues, nor did scientists seek political control. In the second half of the twentieth century, with the development of nuclear weapons, this changed. Scientists served as advisers on political questions, and individual scientists and groups entered political debates. The results have been decidedly mixed. They have brought fresh ideas to the political process, but claims of scientific objectivity have also been used to justify prejudices and narrow academic viewpoints.

An example is the debate over asteroid defense. Scientists have argued since the 1960s that a defense against ballistic missiles was destabilizing because it would cause the Soviets to build their own system, accelerating the arms race. If a U.S. ballistic missile defense was not built, their theory went, the Soviets would show similar restraint. Although the Cold War has passed, the same argument continues to be made against an asteroid impact defense. It is destabilizing, the argument goes, because the capability to deflect an asteroid away from Earth could also be used to deflect one toward Earth. Such an intentional asteroid impact, the argument continued, was more likely than a naturally occurring asteroid impact. Therefore, any efforts to build a defensive system must be stopped.

To deflect an asteroid on a collision course with Earth, it is enough simply to speed it up or slow it down. The asteroid will then cross Earth's orbit at a different time, preventing the collision. Once the minimum Delta-V requirement is met, any additional energy simply increases the miss distance. It would not compromise mission success. Sending an asteroid toward a collision with Earth would require precise changes in its period, inclination, and other orbital characteristics, such that it would not only cross the orbit of Earth but also cross at exactly the same time as Earth at that particular point. It would require considerably more energy and planning to direct an asteroid along such a precise three-dimensional course at a very precise speed. A rock one kilometer or more across would have to be flown with the precision of a space probe on a planetary mission.

On a basic level, such political debates should not enter into scientific questions. The question of whether the impact of a comet or asteroid is sufficient to cause mass extinctions and the question of whether the United States should build a defensive system against a Soviet missile attack are separate issues and should be dealt with as such. The opinion of a scientist on a political issue should be judged, like that of any other individual, on its merits. Given the politicization of scientific questions and the nature of society at the millennium's end, this is not possible.

SCIENCE VS. POLITICS

Despite the bitter political infighting of the impact extinction debate, the two decades of conflict over San Diego streetlighting shows a naivete among scientists in dealing with real-world politics. The question arises as to how the streetlight controversy could happen. The initial impulse is to blame the great social trends of the 1980s and 1990s. The comment of Pam Hamilton, Center City Development Corporation vice president, can serve as a textbook example of the age of narcissism: "People don't feel good about themselves when they don't see color."[9] Yet the actual reasons seem to be in the particular mentality of San Diego.

Kevin Starr, the California state librarian and author of a multivolume history of California, observed that although Pasadena looked "Unto the stars themselves," San Diego was a place "where one might secede from American culture and, edged up against the Pacific, drop out of time and history."[10] While today the rest of California now looks toward the Pacific Rim for its economic future, San Diego feels no need for contact with the outside world and views outsiders with resentment and hostility. This attitude was reflected by one letter writer who said that San Diego should not give up the "safety" of HPS lights "for an installation of any kind 50 miles away."

In retrospect, the astronomers never understood the people they were fighting. The astronomers assumed that if they presented their case factually and logically, the city council would accept it. As intelligent, rational, well-educated people, they could not, and still do not, understand the hatred San Diego feels toward them to this day. For the HPS supporters, the destruction of Palomar Observatory was a benefit. It meant ridding the city of outsiders and closing a window to a larger universe that they felt no connection with.

There is a temptation to dismiss the actions of San Diego as a tale of small-town politics as told by Lewis Carroll or George Orwell, but the events that took place in San Diego mirror those in the larger world. The hostility of San Diego toward astronomy is a reflection of social isolationism. Rather than following a policy of simple noninvolvement with foreign countries, as in the 1930s, San Diego desires to have no connections with anything outside itself.

The trend cuts across all ages and education levels. The *Florida Today* newspaper made an informal survey in 1998 of teenagers at an arcade; one 18-year-old did not know the United States had landed on the Moon. Timothy Ferris, the author of the astronomy book *Coming of Age in the Milky Way,* held a Fourth of July party in 1996. On hand were five college graduate students from Berkeley, Princeton, Brown, Sarah Lawrence, and Wellesley, some of the most prestigious universities in the United States. Ferris asked each one of the students what had happened on the Fourth of July. Not one of them knew.[11]

More significant, these isolationists rage against anything outside their experience, any change, anything different, anything new, and anything that in the slightest way inconveniences them. Theirs is a world where ignorance is superior to knowledge and where curiosity is evil. The blind rage expressed in letters written during the streetlighting controversy is now mirrored every day and night on radio talk shows across the country. It is perfectly fitting that the disgraced ex-mayor of San Diego, Roger Hedgecock, would end up as the host of such a show.

The lessons of the mountains in the sky are many. They show us how our solar system began and how life on Earth can end. The asteroids show us how discoveries are made and the people who make them. They show us the human capability for learning and for folly.[12] The asteroids also show how fragile human existence is, whether from the impact of a billion tons of rock or from human ignorance. Based on the experience of the past 200 years, we can be sure that the third century of asteroid research will bring new discoveries, new questions, and the unexpected among the mountains in the sky.

NOTES

1: DISCOVERY OF THE ASTEROIDS

1. Arthur Berry, *A Short History of Astronomy* (New York: Dover Publications, 1961), 76, 92–99, 125, 131–140, 159–170.
2. Clifford J. Cunningham, *Introduction to Asteroids: The Next Frontier* (Richmond, Va.: William-Bell, 1988), 3, 4.
3. Stanley L. Jaki, "The Titius-Bode Law: A Strange Bicentenary," *Sky and Telescope* (May 1972): 280, 281.
4. Mark Littmann, *Planets Beyond: Discovering the Outer Solar System* (New York: Wiley Science Editions, 1988), 1–17.
5. Clifford J. Cunningham, "The Baron and His Celestial Police," *Sky and Telescope* (March 1988): 271, 272.
6. Clifford J. Cunningham, "Giuseppe Piazzi and the Missing Planet," *Sky and Telescope* (September 1992): 274–275.
7. Cunningham, *Introduction to Asteroids,* 4–7.
8. Lutz D. Schmadel, *Dictionary of Minor Planet Names* (Berlin: Springer-Verlag, 1992), 17.
9. "Pallas Crossing Hercules," *Sky and Telescope* (July 1991): 68; and "The Big Three Asteroids Arrive," *Sky and Telescope* (May 1982): 495.

10. Richard Baum and William Sheehan, *In Search of Planet Vulcan* (New York: Plenum Trade, 1997), 60.
11. Cunningham, *Introduction to Asteroids,* 7, 8, 71; and Charles T. Kowal, *Asteroids: Their Nature and Utilization* (New York: Halsted Press, 1988), 94.
12. Cunningham, *Introduction to Asteroids,* 13.
13. "The Big Three Asteroids Arrive," 497, 498.
14. William Sheehan, *The Planet Mars: A History of Observation and Discovery* (Tucson: University of Arizona Press, 1996), 36–38.

2: VERMIN OF THE SKIES

1. Lutz D. Schmadel, *Dictionary of Minor Planet Names* (Berlin: Springer-Verlag, 1992), 18.
2. Clifford J. Cunningham, *Introduction to Asteroids: The Next Frontier* (Richmond, Va.: William-Bell, 1988), 9. The Rudolph Wolf quotation at the beginning of the chapter is from p. 10.
3. Schmadel, *Dictionary of Minor Planet Names,* 20–26, 39.
4. Joseph Ashbrook, "The Adventures of C. H. F. Peters," *Sky and Telescope* (February–March 1973): 90, 91, 152, 153.
5. Richard Baum and William Sheehan, *In Search of Planet Vulcan* (New York: Plenum Trade, 1997), 185–193.
6. Schmadel, *Dictionary of Minor Planet Names,* 38.
7. Baum and Sheehan, *In Search of Planet Vulcan,* 133–139, 155–169, 178, 194–199, 205–228. Leverrier's Vulcan was named for the Greek god of fire. It should not be confused with the M-class planet of the same name.
8. "Letters," *Sky and Telescope* (September 1980): 190.
9. Cunningham, *Introduction to Asteroids,* 34, 35.
10. Lutz D. Schmadel, Richard M. West, and Claus Madsen, "The Adalberta Mystery," *Sky and Telescope* (January 1983): 33.
11. Cunningham, *Introduction to Asteroids,* 10, 28.
12. "News Notes," *Sky and Telescope* (September 1979): 232.
13. Cunningham, *Introduction to Asteroids,* 10, 11.
14. John E. Bortle, "A Remarkable New England Amateur," *Sky and Telescope* (October 1989): 435, 436.
15. "Notes about Minor Planets," *Sky and Telescope* (September 1977): 193.
16. Joseph Ashbrook, "Astronomical Scrapbook," *Sky and Telescope* (August 1978): 99.
17. Cunningham, *Introduction to Asteroids,* 34, 41, 42, 83.
18. Charles T. Kowal, *Asteroids: Their Nature and Utilization* (New York: Halsted Press, 1988), 22, 23, 58, 59.

19. Cunningham, *Introduction to Asteroids,* 9, 11, 64, 71, 72.
20. William Graves Hoyt, *Lowell and Mars* (Tucson: University of Arizona Press, 1996), 97, 99, 256–258.
21. Ronald Florence, *The Perfect Machine: Building the Palomar Telescope* (New York: Harper Collins, 1994).
22. Cunningham, *Introduction to Asteroids,* 11, 27–29.

3: THE MODERN ERA

1. Clifford J. Cunningham, *Introduction to Asteroids: The Next Frontier* (Richmond, Va.: William-Bell, 1988), 12, 29.
2. "50 and 25 Years Ago," *Sky and Telescope* (May 1996): 8.
3. "3,000 Asteroids and Still Counting," *Sky and Telescope* (November 1984): 412.
4. Charles T. Kowal, *Asteroids: Their Nature and Utilization* (New York: Halsted Press, 1988), 25, 27, 28. The Kowal quotation at the beginning of the chapter is from p. 15.
5. This description is based on the author's experiences as an astronomy student in 1976–1978 at San Diego State University. Things were a whole lot different before PCs, imaging processing software, and CCD cameras.
6. Cunningham, *Introduction to Asteroids,* 63–66.
7. "Rotation of Vesta," *Sky and Telescope* (April 1974): 240.
8. Jeffrey Winters, "Theory Puts Holes in Belief Asteroids Are Solid," *San Diego Union-Tribune* (September 18, 1996), section E; and "Sun Never Sets, for Long, on Fast-Spinning, Water-Rich Asteroid," NASA-JPL press release, July 23, 1999.
9. "The Slowest-Spinning Asteroids," *Sky and Telescope* (June 1983): 504, 505.
10. Cunningham, *Introduction to Asteroids,* 63–67, 77–82.
11. "New Asteroid Data," *Sky and Telescope* (November 1984): 415, 416.
12. "Asteroid Classes: Alphabet Soup," *Sky and Telescope* (April 1983): 326–328.
13. Cunningham, *Introduction to Asteroids,* 54.
14. Richard P. Binzel, M. Antonietta Barucci, Marcello Fulchignoni, "The Origins of the Asteroids," *Scientific American* (October 1991): 93.
15. "Asteroid Classes: Alphabet Soup," 328.
16. "First 'Map' of Vesta," *Sky and Telescope* (December 1983): 502.
17. "Speckled Vesta," *Sky and Telescope* (June 1987): 598.
18. "Hubble Finds Big Crater on Vesta," *Astronomy* (December 1997): 30, 34.
19. Harry Y. McSween Jr., *Meteorites and Their Parent Planets* (Cambridge: Cambridge University Press, 1987), 137–142.
20. "Sing the Asteroid Electric," *Sky and Telescope* (April 1989): 356, 357.
21. Kowal, *Asteroids,* 54.

22. William K. Hartmann and Ron Miller, *Cycles of Fire Stars: Galaxies and the Wonder of Deep Space* (New York: Workman Publishing, 1987), 72–77.
23. McSween, *Meteorites and Their Parent Planets,* 185–188.
24. "Wet Asteroids," *Sky and Telescope* (December 1990): 590.
25. Alan E. Rubin, "Microscopic Astronomy," *Sky and Telescope* (July 1995): 29.
26. "In Search of Ancient Asteroids," *Sky and Telescope* (July 1992): 5, 6.
27. Kowal, *Asteroids,* 54, 57, 58.
28. McSween, *Meteorites and Their Parent Bodies,* 55–57, 84, 187.
29. "Depleting Outer Asteroids," *Sky and Telescope* (June 1997): 18.
30. Alan Stern, "The Sun's Fab Four," *Astronomy* (June 1995): 34–37.
31. Cunningham, *Introduction to Asteroids,* 35, 36.
32. "Lost but Not Forgotten," *Sky and Telescope* (July 1981): 19.
33. "Adalberta Does Not Exist," *Sky and Telescope* (May 1982): 455; and Lutz D. Schmadel, Richard M. West, and Claus Madsen, "The Adalberta Mystery," *Sky and Telescope* (January 1983): 33, 34.
34. "And Then There Were Six," *Sky and Telescope* (January 1983): 34.
35. "And Then There Were Three," *Sky and Telescope* (April 1987): 366–367.
36. "One Down, Two to Go," *Sky and Telescope* (January 1989): 14.
37. "Desperately Seeking Mildred," *Sky and Telescope* (September 1991): 235.
38. Joseph Ashbrook, "Astronomical Scrapbook: An Elusive Asteroid, 719 Albert," *Sky and Telescope* (August 1978): 99, 100.
39. Cunningham, *Introduction to Asteroids,* 2, 89, 90.
40. "532 Herculina as a Double Asteroid," *Sky and Telescope* (September 1978): 210.
41. "Planets and Asteroids That Will Hide Stars in 1981," *Sky and Telescope* (January 1981): 38.
42. Roger W. Sinnott, "Pallas Occultation: A Rare Spectacle," *Sky and Telescope* (May 1983): 440.
43. "Planetary Occultations of Stars in 1984," *Sky and Telescope* (January 1984): 60.
44. "Observers' Notebook," *Sky and Telescope* (March 1980): 261, 262.
45. "Letters," *Sky and Telescope* (November 1980): 372.
46. "Possible Satellite of Asteroid 9 Metis," *Sky and Telescope* (December 1981): 545.
47. "Does 9 Metis Have a Moon?" *Sky and Telescope* (August 1982): 164.
48. "Asteroid Satellite Search," *Sky and Telescope* (February 1983): 193.
49. "Letters," *Sky and Telescope* (February 1984): 108, 109.
50. "Still No Asteroid Satellites," *Sky and Telescope* (November 1987): 455, 456.
51. "Hubble Shoots Vesta," *Sky and Telescope* (April 1995): 14, 15.
52. "The Two Faces of Vesta," *Sky and Telescope* (January 1996): 10.
53. "Newfound Crater Confirms Meteorite Origins," *Sky and Telescope* (November

1997): 18; and "Hubble Finds Big Crater on Vesta," *Astronomy* (December 1997): 30, 34.

4: APOLLOS, AMORS, ATENS, AND CLOSE CALLS: THE NEAR-EARTH ASTEROIDS

1. Brian G. Marsden, "The Recovery of Apollo," *Sky and Telescope* (September 1973): 155.
2. Clifford J. Cunningham, *Introduction to Asteroids: The Next Frontier* (Richmond, Va.: William-Bell, 1988), 93.
3. Marsden, "The Recovery of Apollo," 155, 157.
4. R. L. Duncombe, P. M. Janiczek, and P. K. Seidelmann, "Toro: The Imprisoned Bull?" *Sky and Telescope* (June 1973): 381–383.
5. David Morrison, "Target Earth: It Will Happen," *Sky and Telescope* (March 1990): 264; and MIT Students, *Project Icarus* (Cambridge, Mass: MIT Press, 1979), 1, 2, 4.
6. Marsden, "The Recovery of Apollo," 157.
7. Lutz D. Schmadel, *Dictionary of Minor Planet Names* (Berlin: Springer-Verlag, 1992), 245.
8. Marsden, "The Recovery of Apollo," 158.
9. David Bender, "The Discovery of 1976 AA," NASA History Office, Asteroid Files.
10. "A New Interior Planet," *Sky and Telescope* (March 1976): 158.
11. "Update on 1976 AA," *Sky and Telescope* (April 1976): 230.
12. "New Aten Asteroid," *Sky and Telescope* (July 1986): 30. In contrast to the names from Greek and Roman mythology given to many Apollo-Amor asteroids, Atens receive names from Egyptian mythology.
13. Cunningham, *Introduction to Asteroids,* 93.
14. "Two More Atens," *Sky and Telescope* (March 1990): 251.
15. Marsden, "The Recovery of Apollo," 155–157.
16. "Adonis Recovered," *Sky and Telescope* (April 1977): 243.
17. Schmadel, *Dictionary of Minor Planet Names,* 292, 540, 653; and Morrison, "Target Earth," 264.
18. Lucy McFadden, "The Origin and Composition of Near-Earth Asteroids," NASA History Office, Asteroid Files.
19. John K. Davies, "Is 3200 Phaethon a Dead Comet?" *Sky and Telescope* (October 1985): 317.
20. Cunningham, *Introduction to Asteroids,* 68, 101, 102.
21. "The Frigid World of IRAS-I," *Sky and Telescope* (January 1984): 4–6.
22. John Mason, "Asteroids, Dead Comets, and Meteor Streams," *New Scientist* (December 10, 1988): 34–37.

23. Davies, "Is 3200 Phaethon a Dead Comet?" 317, 318.
24. Mason, "Asteroids, Dead Comets, and Meteor Streams," 37, 38.
25. Gary W. Kronk, *Meteor Showers: A Descriptive Catalog* (Hillside, N.J.: Enslow Publishers, 1988), 3–5, 26, 27, 83, 84, 139, 140, 208, 209; and Lucy-Ann McFadden and Clark R. Chapman, "Interplanetary Fugitives," *Astronomy* (August 1992): 35.
26. "Meteors from Asteroids," *Sky and Telescope* (March 1989): 245, 246.
27. "Asteroid Streams?" *Sky and Telescope* (May 1991): 467, 468.
28. Charles T. Kowal, *Asteroids: Their Nature and Utilization* (New York: Halsted Press, 1988), 73, 74.
29. McFadden, "The Origin and Composition of Near-Earth Asteroids."
30. "An Asteroid with a Tail," *Sky and Telescope* (February 1993): 12, 13.
31. Donald K. Yeomans, *Comets: A Chronological History of Observation, Science, Myth, and Folklore* (New York: John Wiley & Sons, 1991), 352, 353.
32. "Kamikaze Asteroids," *Sky and Telescope* (December 1994): 15.
33. Kowal, *Asteroids,* 46, 47.
34. Andrew J. Butrica, *To See the Unseen: A History of Planetary Radar Astronomy* (Washington, D.C.: NASA, 1996), SP-4218, 6–10, 21–123.
35. "JPL Radar Indicates Icarus Tiny, Very Rough," JPL press release, November 22, 1968.
36. Butrica, *To See the Unseen,* 141, 180, 198–200, 221–224.
37. "Listening to the Solar System 1: Radar Echoes from Asteroids," *Sky and Telescope* (January 1982): 13.
38. Kowal, *Asteroids,* 94, 95.
39. Cunningham, *Introduction to Asteroids,* 21.
40. "NASA Scientists Discover Metal Asteroid near Earth," NASA news release 91-89, June 7, 1991.
41. Butrica, *To See the Unseen,* 248–251.
42. "Radar Imaging Captures Exotic Asteroid," *Sky and Telescope* (January 1990): 13.
43. "Castalia Revealed," *Sky and Telescope* (May 1994): 11.
44. William J. Broad, "NASA Photographs and Asteroid Giving Earth a Close Shave, Sort Of," *New York Times* (January 4, 1993), section B.
45. Alan M. MacRoberts, "The Return of Toutatis," *Sky and Telescope* (December 1996): 76, 77.
46. "A Cosmic Cigar," *Sky and Telescope* (August 1995): 10.
47. Butrica, *To See the Unseen,* 254, 255.
48. Cunningham, *Introduction to Asteroids,* 2, 98, 99.
49. "An Asteroid Whizzes Past Earth," *Sky and Telescope* (July 1989): 30. Had 1989 FC been a C-type asteroid, its size would have been around 430 meters, and the

resulting crater would have been 7.2 kilometers across. The yield would have been 2,300 megatons.

50. "NASA Astronomer Discovers Near-Miss Asteroid That Passed Earth," NASA news release 89-52, April 19, 1989.
51. "Apollo-Catching Made Easy," *Sky and Telescope* (April 1991): 357.
52. Blaine P. Friedlander, "Asteroid Hurled near Earth Jan. 18," *Washington Post* (January 27, 1991); and "Earth Has Close Call with 30-foot Asteroid," *Chicago Tribune* (January 27, 1991).
53. "Mystery Object Buzzes Earth," *Sky and Telescope* (February 1992): 133.
54. Walter N. Webb, "1991 VG: Object Unknown," *MUFON UFO Journal* (April 1992): 15, 16.
55. "Mystery Object Is Just Asteroid," *New York Times* (December 1, 1991).
56. "A Genuine UFO," *Sky and Telescope* (March 1992): 252.
57. "An Asteroid Buzzes Earth," *Sky and Telescope* (September 1993): 9; and Blaine P. Friedlander Jr., "Asteroid's Close Pass Went Undetected," *Washington Post* (June 21, 1993), section A.
58. "The Third Moon-Crossing Asteroid," *Sky and Telescope* (July 1994): 13.
59. "Two Remarkable Asteroids," *Sky and Telescope* (March 1995): 13, 14; and Blaine P. Friedlander Jr., "Asteroid Comes Within 65,000 Miles of Earth," *Washington Post* (December 12, 1994) section A.

5: FAR FRONTIERS: FROM THE TROJANS TO THE KUIPER BELT

1. Charles T. Kowal, *Asteroids: Their Nature and Utilization* (New York: Halsted Press, 1988), 63.
2. Clifford J. Cunningham, *Introduction to Asteroids: The Next Frontier* (Richmond, Va.: William-Bell, 1988), 128, 129.
3. Richard D. Johnson and Charles Holbrow, *Space Settlements: A Design Study* (Washington, D.C.: NASA, 1977), 9.
4. Lutz D. Schmadel, *Dictionary of Minor Planet Names* (Berlin: Springer-Verlag, 1992), 95, 98, 99.
5. Cunningham, *Introduction to Asteroids,* 125, 126.
6. Kowal, *Asteroids,* 64.
7. Cunningham, *Introduction to Asteroids,* 81, 82, 126–128.
8. Clifford J. Cunningham, "The Captive Asteroids," *Astronomy* (June 1992): 43, 44.
9. "Trojan-Type Orbits in the Earth-Sun System," *Sky and Telescope* (June 1974): 373.
10. David H. Levy, *The Quest for Comets* (New York: Plenum Press, 1994), 220, 221.
11. Kowal, *Asteroids,* 64–66.

12. "More about Chiron," *Sky and Telescope* (February 1978): 106; and "News of Chiron," *Sky and Telescope* (July 1978): 5.
13. Cunningham, *Introduction to Asteroids,* 129, 130.
14. "Letters," *Sky and Telescope* (March 1978): 195.
15. "Chiron Brightens," *Sky and Telescope* (April 1988): 358.
16. "Chiron Update," *Sky and Telescope* (June 1988): 583.
17. "2060 Chiron = Comet Kowal?" *Sky and Telescope* (July 1989): 14.
18. Alan Stern, "Chiron Interloper from the Kuiper Disk?" *Astronomy* (August 1994): 29.
19. Richard A. Kerr, "Another Asteroid Has Turned Comet," *Science* (September 2, 1988): 1161.
20. Stern, "Chiron Interloper from the Kuiper Disk?" 29, 30.
21. "Jets on Chiron," *Sky and Telescope* (June 1995): 15.
22. Stern, "Chiron Interloper from the Kuiper Disk?" 30, 31.
23. Alan Stern and Jacqueline Mitton, *Pluto and Charon: Ice Worlds on the Ragged Edge of the Solar System* (New York: John Wiley & Sons,, 1998), 158, 159.
24. Paul R. Weissman, "Comets at the Solar System's Edge," *Sky and Telescope* (January 1993): 27.
25. "Farthest-Ranging Asteroid," *Sky and Telescope* (September 1991): 234, 235.
26. "Outermost Asteroid," *Sky and Telescope* (April 1992): 373; and "New Asteroid Distance Champion," *Astronomy* (June 1992): 22.
27. "An Organic Asteroid?" *Sky and Telescope* (January 1993): 15.
28. Stern and Mitton, *Pluto and Charon,* 158, 159, 162, 163; and Heather Miller, "The Outer Limits," *Air and Space* (April-May 1998): 42, 43.
29. Richard L. S. Taylor, "Deep into That Darkness Peering, Part 1," *Spaceflight* (May 1994): 170, 171. Alas, "Smiley" cannot be used, as asteroid 1613 already has this name, after an orbital calculator and professor, not a fictional master spy.
30. Stuart J. Goldman, "Kuiper Belt Update," *Sky and Telescope* (January 1994): 30.
31. "Pluto's Distant Cousins," *Sky and Telescope* (July 1994): 10.
32. "More Kuiper Comets," *Astronomy* (February 1997): 30; and "More Kuiper Belt Objects," *Astronomy* (December 1997): 30.
33. "This Object Is Far Out," *Astronomy* (August 1997): 24, 26.
34. "Largest Centaur," *Astronomy* (August 1997): 28.
35. "An Oort Cloud Asteroid?" *Sky and Telescope* (August 1997): 19.
36. "Comets on the Edge," *Sky and Telescope* (July 1995): 10.
37. Clyde W. Tombaugh and Patrick Moore, *Out of the Darkness* (Harrisburg, Penn.: Stackpole Books, 1980), 91, 143, 180.
38. Stern and Mitton, *Pluto and Charon,* 145–158, 165–170.
39. "The Kuiper Belt's Dual Colors," *Sky and Telescope* (August 1998): 24.

40. "Astronomers: Kuiper Belt Object from Collision That Created Pluto and Charon," Southwest Research Institute press release, October 12, 1999.

6: ASTEROID SPACE MISSIONS

1. Andrew Wilson, *Solar System Log* (New York: Jane's Publishing, 1987), 22, 23, 27, 28. The Kepler letter at the beginning of the chapter is quoted from Richard S. Lewis, *The Illustrated Encyclopedia of the Universe* (New York: Harmony Books, 1983), 281.
2. Letter, Ernst Stuhlinger to Homer E. Newell, November 21, 1968, NASA History Office, Washington, D.C.
3. Richard O. Fimmel, James Van Allen, Eric Burgess, *Pioneer First to Jupiter, Saturn, and Beyond* (Washington, D.C.: GPO, 1980), 26, 28, 33, 60, 61, 72, 73, 75, 78, 91, 118.
4. Letter, Hannes Alfven to George M. Low, April 10, 1975, NASA History Office, Washington, D.C.
5. Routing slip, T. Burke, April 17, 1975, NASA History Office, Washington, D.C.
6. Letter, Anthony J. Calio to Charles Meyer Jr., NASA History Office, Washington, D.C.
7. "Looking as [sic] Asteroid Multiple Rendezvous Mission," *Defense-Space Daily* (April 22, 1980): 275.
8. William I. McLaughlin, "Space at JPL," *Spaceflight* (September-October 1983): 343.
9. "NASA Looks at Using Modified TIROS for Asteroid Mission," *Defense Daily* (May 27, 1982): 152.
10. "Asteroid Rendezvous Opportunities," *Aerospace Daily* (June 14, 1982): 240.
11. McLaughlin, "Space at JPL."
12. Lutz D. Schmadel, *Dictionary of Minor Planet Names* (Berlin: Springer-Verlag, 1992), 593.
13. "Space Report," *Spaceflight* (April 1984): 154, 155.
14. J. Kelly Beatty, "An Asteroid for the Asking," *Sky and Telescope* (February 1985): 127.
15. William McLaughlin, "Galileo at Venus," *Spaceflight* (March 1989): 93.
16. William McLaughlin, "Galileo Launched," *Spaceflight* (December 1989): 407.
17. "Galileo Mission Threatened," *Spaceflight* (June 1991): 185.
18. "For Extra Credit, Go Get Gaspra," *Sky and Telescope* (February 1992): 135.
19. J. Kelly Beatty, "A Picture-Perfect Asteroid," *Sky and Telescope* (February 1992): 134, 135.
20. "Astronomy Express," *Sky and Telescope* (January 1992): 13.

21. Beatty, "A Picture-Perfect Asteroid."
22. Kathy Sawyer, "First Asteroid with Moon Is Found," *Washington Post* (March 24, 1994), section A.
23. "News Breaks," *Aviation Week and Space Technology* (March 7, 1994): 17.
24. J. Kelly Beatty, "Ida and Company," *Sky and Telescope* (January 1995): 20–22.
25. Dan Durda, "Two by Two They Came," *Astronomy* (January 1995): 33, 34.
26. "Ida Moon Reveals Aged Surface," *Aviation Week and Space Technology* (July 4, 1994): 32.
27. Beatty, "Ida and Company," 22, 23.
28. "New Satellites of Asteroids," *Sky and Telescope* (December 1997): 20; and "Astronomers Discover Moon Orbiting Asteroid," Southwest Research Institute press release, October 7, 1999.
29. Stewart Nozette and Eugene M. Shoemaker, "Clementine Goes Exploring," *Sky and Telescope* (April 1994): 38, 39.
30. Stuart J. Goldman, "Clementine Maps the Moon," *Sky and Telescope* (August 1994): 20–24.
31. Seth Borenstein, "Ice on Moon Could Be Ticket to Expanded Space Travel," *San Diego Union-Tribune* (December 4, 1996), section A.
32. Nozette and Shoemaker, "Clementine Goes Exploring," 39.
33. "Asteroid Flybys That Never Were," *Spaceflight* (September 1994): 312.
34. Goldman, "Clementine Maps the Moon," 20, 21, 24.
35. "NASA Plans New Solar System Missions," *Spaceflight* (May 1993): 151.
36. "NASA Begins Development of First Asteroid-Orbiting Mission," NASA news release 93-220, December 15, 1993.
37. Jim Bell, "Far Journey to a NEAR Asteroid," *Astronomy* (March 1996): 44.
38. Ibid., 44, 45.
39. "What Is 253 Mathilde Like?" Johns Hopkins University NEAR Web Page.
40. "Mission to Asteroid Delayed by Trouble," *San Diego Union-Tribune* (February 17, 1996), section A.
41. "Spacecraft Blasts Off for Asteroid," *San Diego Union-Tribune* (February 18, 1996), section A.
42. "NEAR Fast Approaching Asteroid 253 Mathilde," Johns Hopkins University NEAR Web Page.
43. "Mathilde Images," Johns Hopkins University NEAR Web Page.
44. "Asteroid Mathilde Reveals Her Dark Past," Johns Hopkins University NEAR Web Page.
45. Paul Recer, "Spacecraft's Peek at Asteroid Is Spectacular," *San Diego Union-Tribune* (June 28, 1997), section A.
46. Robert Naeye, "Better Late Than Never," *Astronomy* (April 1999): 24.
47. Paul Recer, "Asteroid Might Have Been Bit of a Planet, Images Show," *San Diego Union-Tribune* (February 18, 2000), section A.

48. "The Mission," DS 1 Home Page, JPL Web Site.
49. "Deep Space 1 Launch Rescheduled to October," JPL press release, April 17, 1998, JPL Web Page.
50. "New Deep Space 1 Trajectory Includes Asteroid Flyby," JPL press release, June 5, 1998; and "Project," DS 1 Home Page, JPL Web Page.
51. "Deep Space 1 Mission Status Report," JPL press release, April 19, 1999.
52. "NASA's Deep Space 1 Mission Target Named after Braille Pioneer," Planetary Society news release, July 27, 1999.
53. "Getting a Feel for Braille," *Sky and Telescope* (October 1999): 28; and "Deep Space 1 Mission Status Report," NASA-JPL press release, July 30, 1999.
54. "Planetary Probes," *Spaceflight* (July 1997): 228.
55. Michael Carroll, "Off-roading on an Asteroid," *Astronomy* (October 1997): 26, 28.
56. "SpaceDev Selects Asteroid Nereus for First Mission," SpaceDev press release, August 26, 1998.
57. "Pluto Stamp Story," Ice and Fire Preprojects, JPL Web Page.
58. "Pluto-Kuiper Express Trajectory Options," Ice and Fire Preprojects, JPL Web Page.
59. Ice and Fire Preprojects, JPL Web Page.

7: THE NAME'S THE THING!

1. Clifford J. Cunningham, *Introduction to Asteroids: The Next Frontier* (Richmond, Va.: William-Bell, 1988), 29.
2. Clifford J. Cunningham, "Giuseppe Piazzi and the Missing Planet," *Sky and Telescope* (September 1992): 275. There was a predecessor for Ceres Ferdinandea. In 1781 William Herschel named the new planet Georgium Sidus (George's Star) after King George III. Where George III did not rule, this was not popular. (The American Revolution was under way at the time.) Although most astronomers used the name Uranus, the British continued to use Herschel's choice until 1850.
3. Lutz D. Schmadel, *Dictionary of Minor Planet Names* (Berlin: Springer-Verlag, 1992), 19.
4. Cunningham, *Introduction to Asteroids,* 31.
5. Schmadel, *Dictionary of Minor Planet Names,* 20, 24–43.
6. It is ironic that one of Luther's asteroids would also carry an allegorical name, 28 Bellona, after the goddess of war, to signify the start of the Crimean War. The name 40 Harmonia was selected to mark the end of the war.
7. Schmadel, *Dictionary of Minor Planet Names,* 6, 44–63, 83.
8. Cunningham, *Introduction to Asteroids,* 27.
9. Schmadel, *Dictionary of Minor Planet Names,* 1–6, 9, 12–14, 645–654.
10. David H. Levy, "Cole of Spyglass Mountain," *Sky and Telescope* (April 1998): 76, 77.

11. Interview with Jennifer Palm, postmaster of Palomar Mountain, September 12, 1997.
12. Schmadel, *Dictionary of Minor Planet Names,* 7.
13. Cunningham, *Introduction to Asteroids,* 33, 34.
14. Schmadel, *Dictionary of Minor Planet Names,* 8.
15. "Astronomy Express," *Sky and Telescope* (July 1990): 14.
16. Schmadel, *Dictionary of Minor Planet Names,* 535, 536, 554, 606. Having spent many long, cold hours at the telescope, the author notes that although classical music captures the glories of the universe, rock and roll keeps you awake.
17. "Crooks, Mad Scientists, and Other Stars," *Washington Times* (April 13, 1990), F2. The writer asked, "How about one for the Gipper?" apparently unaware of the rule that asteroids cannot be named for a political figure, such as President Ronald Reagan, until 100 years after the person's death.
18. Schmadel, *Dictionary of Minor Planet Names,* 13, 14.
19. "Minor Worlds Need Names Too!" *Sky and Telescope* (May 1995): 31.
20. Schmadel, *Dictionary of Minor Planet Names.*
21. David H. Levy, "Some Thoughts on Pet Names for Minor Bodies," *Sky and Telescope* (April 1994): 101, 102; and Letters to the Editor, *Sky and Telescope* (July 1994): 9.

8: 3043 SAN DIEGO: THE UNWANTED HONOR

1. Kevin Starr, *The Dream Endures: California Enters the 1940s* (New York: Oxford University Press, 1997), chapter 3.
2. Neil Morgan, "Who Says We Haven't Made Progress in '94?" *San Diego Union-Tribune* (December 29, 1994), section A.
3. Starr, *The Dream Endures,* chapter 4.
4. Ronald Florence, *The Perfect Machine: Building the Palomar Telescope* (New York: Harper Collins, 1994), 211, 212.
5. Warren Froelich, "Streetlights' Glare Invades Residents' Homes, Lives," *San Diego Union* (August 29, 1983), section B.
6. M. Mitchell Waldrop, "San Diego Picks Sodium over the Stars," *Science* (July 15, 1983).
7. Interview with Michael Anderson, October 11, 1997.
8. Froelich, "Streetlights' Glare."
9. Personal observations of downtown San Diego in the spring of 1984.
10. Froelich, "Streetlights' Glare."
11. Ted Vollmer, "Council Due to Reverse Lights Vote," *Los Angeles Times* (June 1983).
12. "Council Puts Spotlight on Streetlamps Again," undated newspaper clipping.
13. Ibid.

14. Michael Smolens, "Council Reverses Decision to Install Low-Pressure Sodium Streetlights, *San Diego Union* (June 22, 1983), section B.
15. Private source.
16. Warren Froelich, "Streetlights Blind Palomar's Eye to the Universe," *San Diego Union* (August 29, 1983), section B.
17. Linda Kozub, "He Tries to Relight Streetlight Debate," *San Diego Union* (August 11, 1983), section B.
18. Michael L. Ciancone, "David Lasser, Spaceflight Visionary (1902–1996)," *Quest* 5 (No. 2, 1996): 24.
19. Waldrop, "San Diego Picks Sodium."
20. Letters to the Editor, *San Diego Union* (September 17, 1983), section B.
21. Carl Sagan, "Cosmology or Cosmetology?" *San Diego Union* (October 16, 1983), section B. Article reprinted from *Discovery* (October 1983).
22. Froelich, "Streetlights' Glare." Another of Bill Mitchell's more colorful statements was on the issue of fire trucks. In the early 1980s San Diego needed to replace a large number of old fire trucks. The cost would have been considerable, however. Mitchell suggested that the old fire trucks only be sent out on false alarms.
23. "Lights and Safety," *San Diego Tribune* (February 17, 1984), section B.
24. "More than Bug Lights," *San Diego Union* (August 6, 1983), section B.
25. Clipping, *San Diego Tribune* (November 11, 1983), section B.
26. Froelich, "Streelights Blind."
27. Linda Kozub, "Battle over S.D. Streetlights Is Far from Over," *San Diego Union* (November 18, 1983), section B.
28. Private sources.
29. Warren Froelich and Linda Kozub, "Dispute over Street Lights Gains Steam," *San Diego Union* (February 3, 1984), section B.
30. "Residents Invited to Lighting Decision," *San Diego Tribune* (February 3, 1984), section B.
31. "Enormously Significant . . . ," *San Diego Union* (February 3, 1984), section B; and "Let There Be Darkness in Our Sky," *San Diego Tribune* (January 31, 1984), section B.
32. Jeff Ristine, "City Lights: A Bright Day for Science," *San Diego Tribune* (February 7, 1984), section A; and Letter, Curtis Peebles to the Save Palomar Committee, February 11, 1984.
33. "News Notes," *Sky and Telescope* (July 1984): 13.
34. Warren Froelich, "San Diego Gets a Rock in Space," *San Diego Union* (May 2, 1984), section B.
35. Jim Okerblom, "Issue of Lights Treated Lightly in Escondido," *San Diego Union* (July 7, 1984), section B.
36. Claude Walbert, "County Votes to Protect Observatories," *San Diego Tribune* (December 13, 1984), section B.

37. Pat Flynn, "City Plays Party-pooper at Hedgecock's Mansion," *San Diego Union-Tribune* (March 25, 1995), section B.
38. "Lights Discussed," *San Diego Tribune* (February 3, 1984), section B.
39. "Decision on Lighting," *San Diego Tribune* (February 28, 1984), section B.
40. "Starlight Editorial Termed Half-baked," *San Diego Tribune* (December 24, 1984), section B.
41. George Orwell, *1984* (New York: Signet Classic, 1997). Regarding Oceania's astronomy policy: "What are the stars? . . . They are bits of fire a few kilometers away. We could reach them if we wanted to. Or we could blot them out. The Earth is the center of the universe. The Sun and the stars go round it."
42. "Stars-vs.-Security Battle over Lights Heats Up Again," *San Diego Union* (March 27, 1990), section B.
43. "City to Stay with Sodium Street Lights," *San Diego Union* (April 3, 1990), section B.
44. David Graham, "City Lights: Color Them in Dispute," *San Diego Union-Tribune* (March 16, 1992), section B. The *Union* and the *Tribune* had merged the previous year.
45. Clark Dawson, "Clark's Column," *San Diego Astronomy Association Newsletter* (March 1992).
46. Graham, "City Lights."
47. Sandi Dolbee, "In the Name of God," *San Diego Union-Tribune* (November 13, 1994), section D.
48. Clark Dawson, "Clark's Column," *San Diego Astronomy Association Newsletter* (April 1992).
49. David Graham, "Committee Casts Vote for Brighter City Lights," *San Diego Union-Tribune* (March 19, 1992), section B.
50. "Don't Pull the Shade," *San Diego Union-Tribune* (March 18, 1992), section B.
51. Joseph Palca, "Trade a Telescope for Tortellini?" *Science* (April 10, 1992): 167.
52. "Items Infinitum," *San Diego Union-Tribune* (May 3, 1992), section B.
53. "Street Lights: How Many and How Illuminating?" *San Diego Union-Tribune* (March 26, 1992), section B.
54. "Forget the Astronomers, City Lights Should Be Brighter," *San Diego Union-Tribune* (April 5, 1993), section B.
55. Private source.
56. Letter, Ron Roberts to Curtis Peebles, June 8, 1993.
57. Philip J. LaVelle, "City Will Use Whiter Lights," *San Diego Union-Tribune* (September 29, 1993), section A.
58. "Light-Pollution Setback in Southern California," *Sky and Telescope* (December 1993): 8.
59. Peter Rowe, "Who's Happy to Spend $50,000 on Friendliness?" *San Diego Union-Tribune* (July 7, 1996), section D.

60. Philip J. LaVelle, "City to Shed More Light on High-crime Streets South of I-8," *San Diego Union-Tribune* (January 10, 1995), section B.

9: IMPACT

1. Arthur Berry, *A Short History of Astronomy* (New York: Dover, 1961), 149–151.
2. Bevan M. French, *The Moon Book* (New York: Penguin Books, 1977), 37–39, 56–59.
3. John W. Macvey, "Craters of the Moon," *Spaceflight* (January 1961): 13–15.
4. H. H. Nininger, *Out of the Sky: An Introduction to Meteoritics* (New York: Dover, 1952), 205.
5. French, *The Moon Book,* 69.
6. Ronald E. Doel, "The Lunar Volcanism Controversy," *Sky and Telescope* (October 1996): 26–30.
7. Peter Baily, "News Diary," and Patrick Moore, "The Alphonsus Outbreak," *Spaceflight* (January 1959): 11, 21, 23.
8. "Correspondence," *Spaceflight* (May 1961): 107, 108.
9. Macvey, "Craters of the Moon," 14, 15.
10. Walter Alvarez, *T. rex and the Crater of Doom* (Princeton, N.J.: Princeton University Press, 1997), 43–53.
11. "Early Ideas about Impacts and Extinctions (October 1997)," Asteroid and Comet Impact Hazard Web Page.
12. Nininger, *Out of the Sky,* 213, 214.
13. Kathleen Mark, *Meteorite Craters* (Tucson: University of Arizona Press, 1987), 25–37.
14. Peter Lancaster Brown, *Comets, Meteorites, and Men* (New York: Taplinger Publishing, 1973), 188, 189.
15. Mark, *Meteorite Craters,* 43–36, 38, 39.
16. Andrew Chaikin, "Target: Tunguska," *Sky and Telescope* (January 1984): 18–21.
17. Roy A. Gallant, "Journey to Tunguska," *Sky and Telescope* (June 1994): 38, 39.
18. "Tunguska: An Asteroid," *Sky and Telescope* (March 1993): 15.
19. David H. Levy, *The Quest for Comets* (New York: Plenum Press, 1994), 113–118.
20. Interview with Eugene M. Shoemaker, Carolyn Shoemaker, and David H. Levy, May 16, 1996, at Palomar Mountain Lodge and 18-inch telescope.
21. *Asteroids: Deadly Impact,* video, National Geographic Society, 1997.
22. Levy, *The Quest for Comets,* 120, 121, 125–128.
23. French, *The Moon Book,* 132, 133, 246, 247.
24. Alvarez, *T. rex and the Crater of Doom,* 41, 42, 54–56.
25. Robert Jastrow, "The Dinosaur Massacre," *Science Digest* (September 1983): 109.
26. Alvarez, *T. rex and the Crater of Doom,* 59–69.

27. Thomas O'Toble, "Dinosaur Extinction Linked to Collision with Asteroid," *Washington Post* (January 6, 1980), section A.
28. "Early Ideas about Impacts and Extinctions."
29. "The Doomsday Asteroid," *Nova,* PBS, 1995.
30. Boyce Rensberger, "Extinction Theory Challenged," *Washington Post* (April 8, 1985), section A.
31. Walter Sullivan, "Dinosaur Extinction Theory Is Challenged by New Study," *New York Times* (January 28, 1983), section A.
32. Jastrow, "The Dinosaur Massacre," 53.
33. Walter Sullivan, "Fossils Rebut Meteorite as Cause of Dinosaur Death," *New York Times* (January 6, 1982), section A.
34. "More Heat Than Light on the Death of Dinosaurs," *New York Times* (February 6, 1988).
35. Jastrow, "The Dinosaur Massacre," 51, 52, 109. Jastrow's comment that "the problem of the dinosaur extinction has become a non-problem" is similar to the remark that Coon Butte was a "relatively trifling and local affair."
36. Boyce Rensberger, "Are Dissenting Views Being Suppressed," *Science Digest* (March 1986): 32.
37. Paul Hoffman, "Asteroid on Trial," *Science Digest* (June 1982): 63; and Boyce Rensberger, "Death of Dinosaurs: The True Story," *Science Digest* (May 1986): 30.
38. Alvarez, *T. rex and the Crater of Doom,* 80, 81, 96, 101, 102.
39. Philip J. Hilts, "Report Cites Meteorite in Dinosaurs' Extinction," *Washington Post* (November 4, 1983), section A.
40. Rensberger, "Death of Dinosaurs," 31, 32.
41. Wendy S. Wolbach, Roy S. Lewis, and Edward Anders, "Cretaceous Extinction: Evidence for Wildfires and Search for Meteoritic Material," *Science* (October 11, 1985): 167–170.
42. Alvarez, *T. rex and the Crater of Doom,* 86–89.
43. Rensberger, "Are Dissenting Views Being Suppressed," 32.
44. "More Heat Than Light."
45. Alvarez, *T. rex and the Crater of Doom,* 89–97.
46. Boyce Rensberger, "Theory of Dinosaur's Demise Gains Strength," *Washington Post* (May 8, 1987), section A.
47. Alvarez, *T. rex and the Crater of Doom,* 107–112.
48. J. Kelly Beatty, "Killer Crater in the Yucatan?" *Sky and Telescope* (July 1991): 38–40.
49. John Noble Wilford, "For Dinosaur Extinction Theory, a Smoking Gun," *New York Times* (February 7, 1991).
50. Alvarez, *T. rex and the Crater of Doom,* 9–14, 114–122.
51. "A Report of the Discussions at the NASA Advisory Council's New Directions Symposium," June 9–14, 1980, NASA History Office.

52. NASA News, "New Asteroid/Comet Nuclei Hazard Studies Announced," release 86–63, May 20, 1986.
53. David Morrison, "The Spaceguard Survey Report of the NASA International Near-Earth-Object Detection Workshop," Jet Propulsion Laboratory, January 25, 1992. The AIAA is the successor organization to the American Rocket Society, founded by David Lasser. Asteroid 1989 FC was eventually named 4581 Asclepius. Although the Spaceguard proposal was not the cause of the controversy over the Los Alamos meeting, it was attacked by a writer with the *Washington Times*. In an editorial titled "The Asteroids Are Coming. Not!" published on April Fools Day 1992, Spaceguard was called "this latest and boldest scam to make away with the taxpayers' money." Repeatedly the term "stargazers" was used as a pejorative. It said of the impact extinction theory, "For some years now, the scientificoes have been weaving a yarn that what killed the dinosaurs was just such an asteroid."
54. Robert L. Park, "Star Warriors on Sky Patrol," *New York Times* (March 25, 1992), section A.
55. Joel Achenbach, "Why Things Are Heads Up," *Washington Post* (February 12, 1993), section C. Achenbach's suspicions are not valid. The interest in asteroid defenses dated to 1984, at the depths of the Cold War. Congressional interest was triggered by the near-miss of 1989 FC in March 1989. This was more than two and a half years before the fall of the USSR. The congressional directive to hold the workshops was made in September 1990, fifteen months before the fall of the USSR.
56. Fran Smith, "Star Dreck," *West* (March 22, 1992).
57. Kathy Sawyer, "Shooting Back at Space Rocks?" *Washington Post* (March 30, 1992), section A.
58. Leonard David, "Defense Experts Duck Asteroid Threat Hearing," *Space News* (March 29–April 4, 1993): 6.

10: SHOEMAKER-LEVY 9

1. Interview with Eugene M. Shoemaker, Carolyn Shoemaker, and David H. Levy, May 16, 1996, at Palomar Mountain Lodge and 18-inch telescope.
2. David H. Levy, *The Quest for Comets* (New York: Plenum Press, 1994), 147–151.
3. Ronald Florence, *The Perfect Machine: Building the Palomar Telescope* (New York: Harper Collins, 1994), 151, 152, 260, 261, 304, 305, 349, 351.
4. David H. Levy, "Star Trails," *Sky and Telescope* (June 1991): 658, 659.
5. Interview with David H. Levy, May 16, 1996, at Palomar 18-inch telescope.
6. Duncan Steel, "Project Spaceguard: The Fiction and the Fact," *Spaceflight* (July 1993): 219–223.

7. Levy, *The Quest for Comets,* 224, 225.
8. David H. Levy, *Impact Jupiter* (New York: Plenum Press, 1995), chapter 2; and David H. Levy, "Pearls on a String," *Sky and Telescope* (July 1993): 38, 39.
9. "Comets on a String," *Sky and Telescope* (June 1993): 8.
10. Levy, *Impact Jupiter,* 7, 27, 38–41, 46–50.
11. J. Kelly Beatty and David H. Levy, "Awaiting the Crash," *Sky and Telescope* (January 1994): 41–43.
12. J. Kelly Beatty and David Levy, "Awaiting the Crash Part II," *Sky and Telescope* (July 1994): 18–23.
13. Robert Burnham, "Behind the Scenes: 1994's Great Comet Crash," *Astronomy* (October 1994): 6.
14. Levy, *Impact Jupiter,* 86–88, 105–109.
15. Beatty and Levy, "Awaiting the Crash Part II," 20.
16. Levy, *Impact Jupiter,* 128, 139, 140.
17. "When the Comet Fragments Will Hit," *San Diego Union-Tribune* (July 13, 1994), section E.
18. Levy, *Impact Jupiter,* 141, 146, 147–155.
19. "The Doomsday Asteroid," *Nova,* PBS, 1995.
20. David Graham, "Comet Jolts Jupiter: It's Not a Dud," *San Diego Union-Tribune* (July 17, 1994), section A.
21. Alan M. MacRobert, "Amateur Astronomy's Greatest Week," *Sky and Telescope* (October 1994): 25.
22. Stephen James O'Meara, "The Great Dark Spots of Jupiter," *Sky and Telescope* (November 1994): 32, 33, 35.
23. J. Kelly Beatty and David H. Levy, "Crashes to Ashes: A Comet's Demise," *Sky and Telescope* (October 1995): 19.
24. Levy, *Impact Jupiter,* 163–176; and David Graham, "Awesome Impacts Blast Giant Jupiter," *San Diego Union-Tribune* (July 19, 1994), section A.
25. "Comet's Impacts Dwarf Earth's Nuclear Arms," *San Diego Union-Tribune* (July 19, 1994), section A.
26. O'Meara, "The Great Dark Spots of Jupiter," 32.
27. Beatty and Levy, "Crashes to Ashes," 19, 20; and Levy, *Impact Jupiter,* 196–198.
28. O'Meara, "The Great Dark Spots of Jupiter," 35.
29. Beatty and Levy, "Crashes to Ashes," 20, 21, 24–26.
30. Levy, *Impact Jupiter,* 184, 185.
31. Burnham, "Behind the Scenes," 6.
32. O'Meara, "The Great Dark Spots of Jupiter," 32, 33.
33. Beatty and Levy, "Crashes to Ashes," 25.
34. Interview with Shoemaker, Shoemaker, and Levy.

35. Rex Graham, "Making an Exceptional Impact," *Astronomy* (May 1998): 37–39.
36. David E. Graham, "Astronomer's Ashes Carried on Moon Probe," *San Diego Union-Tribune* (January 7, 1998), section A; and "Lunar Prospector's Flashy Finale," *Sky and Telescope* (October 1999): 29.
37. Interview with Shoemaker, Shoemaker, and Levy.
38. "NEAT Begins Operation," NASA-JPL press release, April 24, 1996, JPL Web Page.
39. "General Information: Near-Earth Asteroid Tracking (NEAT) Page," JPL Home Page.
40. "Astronomers Find New Class of Asteroid," University of Hawaii news release, July 1, 1998.
41. IAU Minor Planets Center Web Page.
42. "Amateur Finds Aten Asteroid," *Sky and Telescope* (September 1997): 17.
43. "News Archive 1998: New NASA Office to Focus on Asteroid Detection," Asteroid and Comet Impact Hazards Web Page.
44. "The LINEAR Project" and "History of Lincoln Near Earth Asteroid Research (LINEAR) Project," MIT-Lincoln Laboratory Web Page.
45. Tony Ortega, "Air Force Telescope Preempts Comet Hunters," *Astronomy* (April 1999): 60, 61.
46. "News Archive 1999: The NASA-USAF NEO Search Program," Asteroid and Comet Impact Hazards Web Page.
47. "Asteroid Hunters Bring Oldie-but-Goodie Telescope into New Age," NASA-JPL news release, June 24, 1999.
48. J. Kelly Beatty, "Secret Impacts Revealed," *Sky and Telescope* (February 1994): 26, 27.
49. Congressional Testimony, "Statement of John D. G. Rather," Asteroid and Comet Impact Hazards Web Page.
50. William J. Broad, "Military Satellites Prove Earth Bombarded by Meteoroids," *San Diego Union-Tribune* (February 16, 1994), section E.
51. Beatty, "Secret Impacts Revealed," 27.
52. "Fishermen Report an Astronomical Whopper," *San Diego Union-Tribune* (May 1, 1994), section A.
53. "Satellites Detect Record Meteor," *Sky and Telescope* (June 1994): 11.
54. "When It Comes to Detecting Meteors, Los Alamos Researcher Is All Ears," Los Alamos National Laboratory press release 96-201, December 19, 1996.
55. "Los Alamos Array Detects Large, Bright Meteor: Laboratory Researcher Joins the Search," Los Alamos National Laboratory press release 97-155, October 10, 1997.
56. USAF Bolide Information Releases, Headquarters Air Force Technical Applications Center Web Page.

57. William J. Broad, "Earth Is Target for Rocks at Higher Rate Than Thought," *New York Times* (January 7, 1997).

11: PLANETARY DEFENSE

1. "Early Ideas about Impacts and Extinctions," Asteroid and Comet Impact Hazard Web Page (October 1997).
2. MIT Students, *Project Icarus* (Cambridge, Mass.: MIT Press, 1979) 3–69, 155–160.
3. "Systems Engineering Avoiding an Asteroid," *Time* (June 14, 1967): 54, 55.
4. MIT Students, *Project Icarus,* 18–23, 32, 37, 38, 74–78, 121–128, 148–151.
5. Roger E. Bilstein, *Stages to Saturn: A Technological History of the Apollo-Saturn Launch Vehicles* (Washington, D.C.: NASA, 1980), 209–233, 351–363.
6. Donald K. "Deke" Slayton with Michael Cassutt, *Deke! U.S. Manned Space: From Mercury to the Shuttle* (New York: Forge, 1994), 208–210.
7. Bilstein, *Stages to Saturn,* 362, 363.
8. "Our Awesome Atomic Arsenal," *Reader's Digest* (November 1963): 69.
9. Yefim Gordon and Vladimir Rigmant, *Tupolev Tu-95-142 Bear* (North Branch, Minn.: Aerofax, 1997), 45–47.
10. F. Kasman et al., "Proton: Development of a Russian Launch Vehicle," *Journal of the British Interplanetary Society* (January 1998): 3, 4.
11. Rob R. Landis, "The N-1 and the Soviet Manned Lunar Landing Program," *Quest* (Winter 1992): 21–30.
12. Congressional Testimony, "Statement of John D. G. Rather," Asteroid and Comet Impact Hazards Web Page.
13. Peter Tyson, "Comet Busters," *Technology Review* (February-March 1995): 25, 26.
14. Gerrit L. Verschuur, "Impact Hazards: Truth and Consequences," *Sky and Telescope* (June 1998): 29–31.
15. John Noble Wilford, "What If a Huge Asteroid Hits the Atlantic?" *New York Times* (January 8, 1998).
16. James V. Scotti, "The Tale of an Asteroid," *Sky and Telescope* (July 1998): 30–34.
17. "News Archive 1998: Original IAU Release, March 11, 1998," Asteroid and Comet Hazards Web Page.
18. Stuart J. Goldman, "The Most Dangerous Rocks in Space," *Sky and Telescope* (June 1998): 33.
19. Adam Scott and Sharon Begley, "Never Mind!" *Newsweek* (March 23, 1998).
20. Malcolm W. Browne, "Old Photos Helped Refine Progress of Asteroid," *New York Times* (March 14, 1998).
21. "Amateurs Alleviate Asteroid Threat," *Sky and Telescope* (October 1999): 17.
22. J. Kelly Beatty, "The Torino Scale: Gauging the Impact Threat," *Sky and Telescope* (October 1999): 32, 33; and "MIT Researcher Creates Scale to Assess Earth-Asteroid Close Encounters," *MIT News,* July 22, 1999.

23. Congressional Testimony, "Statement of John D. G. Rather," Asteroid and Comet Impact Hazards Web Page.
24. Scotti, "The Tale of an Asteroid," *Sky and Telescope* (July 1998): 33, 34.
25. John Boudreau, "Collision Course," *Washington Post* (April 6, 1994), section C.
26. Wayne R. Austerman, "Program 437: The Air Force's First Antisatellite System" (Peterson AFB: Air Force Space Command, April 1991), 13, 14.

12: THE THIRD CENTURY OF ASTEROID STUDIES

1. Clifford J. Cunningham, "Giuseppe Piazzi and the Missing Planet," *Sky and Telescope* (September 1992): 274–275.
2. Clifford J. Cunningham, "The Baron and His Celestial Police," *Sky and Telescope* (March 1988): 271, 272.
3. Interview with Eugene M. Shoemaker, Carolyn Shoemaker, and David H. Levy, May 16, 1996, at Palomar Mountain Lodge and 18-inch telescope; and *Asteroids: Deadly Impact,* video, National Geographic Society, 1997.
4. Walter Alvarez, *T. rex and the Crater of Doom* (Princeton, N.J.: Princeton University Press, 1997).
5. "Amateur Finds Aten Asteroid," *Sky and Telescope* (September 1997): 17.
6. Alvarez, *T. rex and the Crater of Doom.*
7. William Sheehan and Thomas Dobbins, "Le Verrier's Wild Geese," *Sky and Telescope* (October 1998): 112–115.
8. See, for example, G. J. H. McCall, *Meteorites and Their Origins* (New York: John Wiley & Sons, 1973).
9. Neil Morgan, "Here, You Take Over; I'm Going for a Walk," *San Diego Union-Tribune* (April 22, 1993), section A.
10. Kevin Starr, *The Dream Endures: California Enters the 1940s* (New York: Oxford University Press, 1997), 287, 288.
11. Billy Cox, "A Heritage Lost in Space?" *Florida Today* (August 23, 1998).
12. On May 1, 2000, the last missing numbered asteroid, 719 Albert, was spotted by the Spacewatch telescope. The identification was made by Gareth V. Williams at the Minor Planet Center. Looking at the orbit, Williams thought it was similar to that of Albert, but had to spend considerable time linking together the 1911 and 2000 observations. Brian Marsden reviewed his calculations and concurred that Albert had finally been recovered after 89 years.

INDEX